8/07

gineers:
A Student's
Handbook

D1630470

Maths for Engineers: A Student's Handbook

RUTH REES and
MAYVEEN BLACKWELL

A member of the Hodder Headline Group
LONDON • SYDNEY • AUCKLAND

First published in Great Britain 1999 by Arnold,
a member of the Hodder Headline Group,
338 Euston Road, London NW1 3BH

http://www.arnoldpublishers.com

British Library Cataloguing in Publication Data
A catalogue record for this book is available from the British Library

ISBN 0 340 70586 8

1 2 3 4 5 6 7 8 9 10

Commissioning Editor: Sian Jones
Production Editor: Julie Delf
Production Controller: Sarah Kett
Cover design: Mouse Mat

Typeset by AFS Image Setters, Glasgow
Printed and bound in Great Britain by
J W Arrowsmith Ltd, Bristol

What do you think about this book? Or any other Arnold title?
Please send your comments to feedback.arnold@hodder.co.uk

Contents

Preface and acknowledgements

This book arose from the concern expressed by lecturers and bodies, such as the Engineering Council, over the poor mathematical performance of engineering students. We hope that *Maths for Engineers: A Student's Handbook* will contribute towards a better understanding by students.

Embedded in this book are the results both of many years of research into students' mathematical difficulties and considerable experience of the teaching of applied mathematics.

We have appreciated our many discussions with lecturers who so willingly gave their time.

Ruth Rees, in particular, wishes to thank her huband, Dr John T. Thomas, for his apparently inexhaustible patience and stamina in the correction of some very difficult proofs.

chapter

1 An introduction to mathematics

Why study maths?

If you were asked this question, what would you say? 'I've got to pass an exam in maths to get a job' or 'It helps us do the shopping, budget and even work out our income tax'. You might add that 'some of my friends have had to resit GCSE maths several times!'

Mathematics seems to bother a lot of people and appears to be a difficult hurdle in school and college. The question 'Do you **like** maths?' can produce a mixture of responses, some rude!

So why study maths? The answer is that mathematics, because of its nature, is essential to our everyday lives, studies and work. Maths is a powerful language because it is so concise and generalizable. Think of $E = mc^2$. Do you know what it means? These symbols and signs are part of the language and are a very effective shorthand for engineers. You could not study shape and space without a knowledge of algebra and you cannot solve problems without a good grasp of numbers.

1.2 Maths in this book

The secret of a strong building lies in its foundation. So it is with your engineering course: success depends on your understanding of maths, which is the foundation of engineering. All the maths in this book will be useful to you not only in engineering, but also in everyday life and in whatever other courses and work you undertake. There are many examples throughout the book of applications to engineering and exercises for you to work through. With practice, you will acquire the necessary skills. There are also examples in other contexts because one of the secrets of true learning is to be able to apply mathematics in many different situations. Mathematical skills in engineering, commerce and economics are essentially the same in kind, although how much you will use clearly depends on the context.

There are 8 chapters in the book, of which the first few, for some of you, may act as a 'refresher' course. Do not skip these chapters because a clear understanding of number and how we use the rules is important for a clear understanding of algebra and all that follows. For other students these first chapters may be essential reading as a necessary preparation for the course.

Mathematics can be thought of as a language using symbols and signs as described in Chapter 5. There also are other important ways of representing mathematics such as **visually** in the form of tables, charts and graphs. In these, data are classified and represented in concise ways which help us to interpret information, and in some cases forecast possible future trends, as in Chapter 4. Graphs are dealt with in detail in Chapter 7 and you are shown how to interpret experimental results in Chapter 8.

As you know, when applying mathematics to your engineering tasks it is essential to understand accuracy i.e. how many significant figures or decimal places are required in your solutions. You will be dealing with physical quantities so you will need a good understanding of units. Relationships between physical quantities and the laws which govern their behaviour are covered.

Summaries are provided in most chapters to help you recall the key points. The many exercises give you the opportunity to practise and develop your skills in all those aspects of mathematics you need at this level of your studies.

Units are an important part of your work and at the end of the book you will find the section on the SI system to be a useful reference. You will note that, in the text, the unit for velocity, for example, is expressed as m/s rather than as $m\,s^{-1}$. As you progress in your studies you will realize the advantage of using negative indices.

1.3   Maths and you: how much do you know?

You know that to keep physically fit you must exercise on a regular basis. So it is with maths. You can become competent and **agile** with numbers while having lots of fun by exploring the ways in which numbers behave and form patterns. If you develop a real 'feeling' for and agility with numbers you will find it relatively easy to move to **generalizing number**, in algebra and in the many techniques necessary for problem-solving in engineering.

REMEMBER: **Maths is here to make life simple!**

To find out just where you are you can turn to the self-assessment test at the back of this book. Have a go! Keep your score (answers also at the back of the book). Spot any weaknesses you may have and turn to the appropriate chapter in the book for help.

In a couple of months, or sooner, have another go and compare your scores. We, the writers of this book, wish you every success.

chapter

2

Using whole and mixed numbers

The decimal system

This is the number system in general use having a base of ten. In this system zero is used together with the natural numbers so that we have the digits 0, 1, 2, ... , 9. All other numbers are written as combinations of these digits and the invention of zero not only enables us to express **nothing** but to also make numbers ten times (or multiples of ten) larger or smaller.

The numbers are placed in columns of units, tens, hundreds and thousands (if any). Units are single numbers, i.e. less than ten.

U   units
T   tens
H   hundreds
Th   thousands

It is important to write the digits in the correct columns.

EXAMPLES

Table 2.1 gives some examples.

Table 2.1 Using columns to show place value

Words	Digits			
	Th 1000	H 100	T 10	U 1
Forty nine			4	9
Two hundred and seventy three		2	7	3
Three thousand and ninety	3	0	9	0
One thousand and one	1	0	0	1
Six thousand and eleven	6	0	1	1

Carelessness can considerably alter the value of the number. Familiarity with this **place value** of the system is essential.

EXERCISE 2.1

State what each of the digits represents in

(a) 101 267   (b) 39 050

In order to see the relative size and order of numbers it is helpful to represent them by points on a line. A **number line** can be any length since there is no end to the natural numbers. If you started to count now you would never finish in your lifetime!

EXAMPLE

Mark the number 55 on both number lines in Fig. 2.1.

Figure 2.1 The number 55 on two number lines

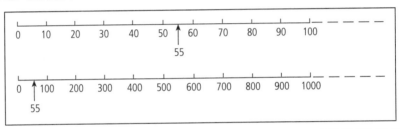

You can see that both number lines give the relative position of 55 but that it is more accurately located on the first line.

2.2 Introduction to common fractions and decimal fractions

In practical situations involving measurement we are usually working with numbers that are whole numbers plus parts of whole numbers, for example 3.5 or $3\frac{1}{2}$. We shall call these **mixed** numbers. The parts of a whole, for example 0.5 or $\frac{1}{2}$, are called **fractions** and there are two kinds, common fractions and decimal fractions.

Common fractions are those used in everyday language such as $\frac{1}{2}$ (one half), $\frac{3}{4}$ (three quarters) and so on. **Decimal fractions** such as 0.5 (zero point five) 0.75 (zero point seven five) are used more in scientific work because, being part of the place value system, they are more easily compared. Common fractions and decimal fractions are actually different forms of the **same number**. For example 0.5 is $\frac{1}{2}$ and 0.75 is $\frac{3}{4}$; which form we use in any situation depends on the context.

In your engineering work you will be using mostly decimal fractions but common fractions will also occur so it is important that you understand the close connection between them and how to operate with them. For example, if you are calculating the effective resistance of two or more resistors in parallel (as you will later in your studies), then you are involved in the addition of common fractions. Sometimes it is quicker to solve a problem using common fractions than it is to convert decimal fractions using a calculator.

Decimal fractions, as the name suggests, are simply common fractions expressed as tenths, hundredths, thousandths but written using a decimal point instead of the denominator 10 and multiples of ten. For example, $\frac{6}{10}$, $\frac{6}{100}$ . . ., are written 0.6, 0.06, ... This way of writing decimal fractions is shown in Table 2.2.

Table 2.2 Decimal fractions

Words	Tens T 10s	Units U 1s	Decimal point	Tenths t $\frac{1}{10}$ ths	Hundredths h $\frac{1}{100}$ ths
Zero point one or one tenth		0	.	1	
Zero point one five or 15 hundredths		0	.	1	5
Twenty-three point four nine or twenty-three and forty-nine hundredths	2	3	.	4	9

Take careful note of the following:
- 0.1 is the same as .1. The zero is used to show that there is **no** whole number.
- 0.1 is the same as 0.10 or 0.100, etc. Zero is used in this way to emphasize no tenths, no hundredths, etc. It is also a useful way of expressing degrees of accuracy in experimental measurement.
- 0.1 is the same as $\frac{1}{10}$ OR $\frac{10}{100}$ OR $\frac{100}{1000}$, etc.
- 0.15 means $\frac{1}{10} + \frac{5}{100}$ OR $\frac{15}{100}$.
- 23.49 means $20 + 3 + \frac{4}{10} + \frac{9}{100}$.

Test your understanding with the following exercise.

EXERCISE 2.2

Copy and complete Table 2.3 by filling in the words or digits as appropriate

---

EXERCISE 2.3 Place value

In the following numbers, state what each digit represents.

   1. 47.63   2. 108.69

   3. 0.102   4. 1012.69

   5. 618.20

Table 2.3 Decimal fractions in words and numbers

Words		Digits				
		T	U	.	t	h
1. Seventeen and two tenths						
2. Seventy-one and fourteen hundredths						
3. Three and twenty hundredths						
4. Five and seven hundredths						
5. One hundredth						
6.		1	1	.	2	0
7.			4	.	0	7
8.		1	0	.	0	3
9.			0	.	6	9
10.			0	.	0	5

As with the natural numbers, the number line can also help to establish the relative sizes of numbers.

Figure 2.2 Decimals on the number line

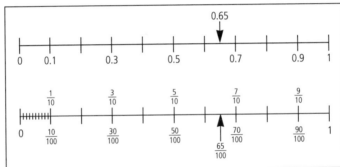

You will see from Fig. 2.2 that even between zero and one there is an infinity of points and therefore numbers. You would need more than nine lives to count these and you would still not get to the end!

In order to compare the relative sizes of numbers, use zero as in the following example.

EXAMPLE

Arrange the following numbers in order of size beginning with the smallest:
0.33   0.25   0.7   0.69.

Write 0.7 as 0.70 for easy comparison.

Then the order is: 0.25   0.33   0.69   0.70.

EXERCISE 2.4 Relative size of decimal fractions

Arrange the following numbers in order of size beginning with the smallest:

1. 0.88   0.8   0.81   0.9   0.89

2. 0.43   0.5   1.05   0.0450   0.495

2.3 Arithmetic operations

We call addition, subtraction, multiplication and division **operations**. You will already be familiar with these.

There are certain **rules** to obey to ensure that your calculations are correctly carried out. These rules also form the basis of your work in algebra.

- The order of addition does **not** matter.
  For example, $3 + 5 = 5 + 3$.
- The order of subtraction **does** matter.
  For example, $5 - 3 \neq 3 - 5$ (note that $\neq$ means '**not** equal to').
- The order of multiplication does **not** matter.
  For example, $5 \times 3 = 3 \times 5$.
  This fact is extremely useful since it is much easier to carry out, say, 3×528 than 528×3 (3 lots of 528 compared with 528 lots of 3).
- The order of division **does** matter.
  For example, $528 \div 3 \neq 3 \div 528$.

The **order** in which these operations are carried out matters very much. Before you begin to use a calculator you must become proficient in using operations in the correct order. The rule is: 'multiplication and division must always be carried out **before** addition and subtraction.' For example,

$$5 + 7 \times 12 \div 4 = 5 + 7 \times 3 \quad \text{OR} \quad 5 + 7 \times 12 \div 4 = 5 + 84 \div 4$$
$$= 5 + 21 \qquad\qquad\qquad\qquad = 5 + 21$$
$$= 26 \qquad\qquad\qquad\qquad\quad = 26$$

- There is an important addition to this rule when **brackets** are used to group together some of the numbers and their operators. For example, 'three lots of (two plus one)' is not the same as 'three lots of two, plus one'. In the first case we write $3(2+1)$ but in the second case we write $3 \times 2 + 1$.

Brackets have top priority in the order of operations and the general rule can be remembered using the acronym **BODMAS**. The letters stand for Brackets Of Division Multiplication Addition Subtraction. For example,

$$(5 \times 7) + 4 \div (10 - 8) = 35 + 4 \div 2$$
$$= 35 + 2$$
$$= 37$$

Note the BODMAS order carefully. **Brackets must be worked first.**

'Of' means multiply.

2.4 Operations with decimal fractions

Operations of $+ \; - \; \times \; \div$ on **mixed** numbers obey the same rules as the whole numbers.

Addition and subtraction

Addition and subtraction for decimal fractions are straightforward provided you understand the decimal place value system. You do not always need a calculator! For example if you wish to add 0.12 and 0.5 then setting it out quickly like this with the digits in the right columns means that you should not go wrong.

$$\begin{array}{r} 0.12 \\ 0.50 \\ \hline 0.62 \end{array}$$

Note that 0.5 has been written as 0.50 to help line up the correct 'columns' and decimal point.

It is in multiplication and division that mistakes commonly occur. In this section, therefore, we are going to emphasize the effect of the operators × ÷ on decimal fractions.

Multiplication

There are indeed surprises. Study the following examples carefully. You may find it helpful to think of × as 'of'.

EXAMPLES

$0.9 \times 0.9 = 0.81$	OR	$\dfrac{9}{10} \times \dfrac{9}{10} = \dfrac{81}{100}$
$0.8 \times 0.7 = 0.56$	OR	$\dfrac{8}{10} \times \dfrac{7}{10} = \dfrac{56}{100}$
$0.3 \times 0.3 = 0.09$	OR	$\dfrac{3}{10} \times \dfrac{3}{10} = \dfrac{9}{100}$
$0.1 \times 0.01 = 0.001$	OR	$\dfrac{1}{10} \times \dfrac{1}{100} = \dfrac{1}{1000}$

Could you have done these without a calculator?

A helpful rule is to multiply the numbers as if they were whole and then put in the total number of decimal places. So in the first three answers 2 d.p. are inserted, whilst in the fourth answer we need 3 d.p.

Are you surprised at the answers? Do you observe that each answer is **smaller** than either number in the question even though we have multiplied? So, 'a **part of a whole one of (×) a part of a whole one' must be less than either part**. For example, 0.56 is **smaller** than either 0.8 or 0.7. Thus the effect of multiplying two decimal fractions or two common fractions (as in $\frac{9}{10} \times \frac{9}{10} = \frac{81}{100}$ etc.) is to **decrease** the answer rather than increase as with multiplication of whole or mixed numbers.

What of the example in between, i.e. with one number a part of a whole one? Consider 0.2×19. This means 0.2 lots of 19, which is 3.8 and certainly less than 19. Thus **multiplication by a number less than one makes smaller.**

Division

Remember that division is the inverse of multiplication and therefore in effect we 'turn the divisor upside down and multiply'. Look at the following examples carefully to see the effect of division on two decimal fractions.

To divide we either change the divisor to a whole number by multiplying top and bottom OR we change to tenths, then invert and multiply. With practice you can do them 'in your head' but you should always check your answer.

The first example is written in full.

EXAMPLES

Change divisor to whole number   OR   Change to tenths, invert and multiply

1. $0.56 \div 0.8 = \dfrac{0.56}{0.8} \times \dfrac{10}{10}$ OR $= \dfrac{56}{100} \div \dfrac{8}{10}$

$ = \dfrac{5.6}{8}$ $ = \dfrac{56}{100} \times \dfrac{10}{8}$

$ = 0.7$ $ = \dfrac{7}{10}$

$ = 0.7$

CHECK: division and multiplication are inverse operations. Thus 0.7×0.8 should give us 0.56 which it does.

Look through the following examples carefully.

2. $0.9 \div 0.3 = 3$ OR $\dfrac{9}{10} \times \dfrac{10}{3} = 3$

3. $0.6 \div 0.3 = 2$ OR $\dfrac{6}{10} \times \dfrac{10}{3} = 2$

4. $0.1 \div 0.1 = 1$ OR $\dfrac{1}{10} \times \dfrac{10}{1} = 1$

5. $0.2 \div 0.4 = 0.5$ OR $\dfrac{2}{10} \times \dfrac{10}{4} = \dfrac{1}{2}$

In example 4 above you see that a number divided by itself is one. This is true of any number, even decimal fractions!

What else do you observe in these examples? You should note that, firstly, all the answers are **bigger** than the number you started with e.g. $0.7 > 0.56$. Are you surprised?

Secondly, the answers to examples 2 and 3 are even bigger than 1. Think what the $\div$ sign is asking – 'How many 0.3's in 0.9?' It's not so surprising that the answer is 3 whole ones. The divisors 0.8, 0.3, 0.1 and 0.4 are each less than one and so we should expect the answers to be **larger** than 0.56, 0.9, 0.6, 0.1 and 0.2. Thus in effect **division by a number less than one makes bigger**.

EXERCISE 2.5 Multiplication of decimal fractions

You should be able to do the following questions quickly. Work through them without a calculator and then check with your calculator.

[HINT: Multiply as whole numbers. Then put in the number of d.p. or change to tenths.]

(a) 0.9×0.8 (b) 0.7×0.7
(c) 0.5×0.5 (d) 0.3×0.2
(e) 0.2×0.1 (f) 0.02×0.3
(g) 0.03×0.04 (h) $0.9 \times 0.1 \times 0.2$
(i) 0.2×25 (j) 3×0.3

EXERCISE 2.6 Division of decimal fractions

[HINT: Divide by changing divisor to whole number by multiplying top and bottom OR change to tenths, invert and multiply.]

(a) $0.8 \div 0.4$ (b) $0.4 \div 0.2$
(c) $0.3 \div 0.15$ (d) $0.25 \div 0.5$
(e) $0.3 \div 0.6$ (f) $0.09 \div 0.3$
(g) $0.01 \div 0.1$ (h) $1 \div 0.5$
(i) $10 \div 0.2$ (j) $50 \div 0.1$

2.5   Operations with common fractions

Relative sizes of common fractions

Students often have difficulty in seeing quickly which common fractions are larger or smaller than others. So first acquire some agility with common fractions.

You will realize from experience that $\frac{1}{2} > \frac{1}{4}$, but do you realize that $\frac{1}{5} > \frac{1}{6}$? Can you see that the common fractions $\frac{1}{9}, \frac{1}{7}, \frac{1}{4}, \frac{1}{3}, \frac{1}{2}$, have been arranged in order of size starting with the smallest? If you think carefully you will see that the bigger the divisor the smaller must be the fraction.

Decimal fractions are much easier to compare. If we convert the above common fractions to decimal fractions using the calculator, e.g. $1 \boxed{\div} 9 \boxed{=} 0.11$ (2 d.p.), we get 0.11, 0.14, 0.25, 0.33, 0.50 and these are in ascending order of size. Try the following for yourself.

EXERCISE 2.7 Relative sizes of fractions

1. Arrange the common fractions $\frac{1}{6}, \frac{1}{9}, \frac{1}{4}, \frac{1}{7}, \frac{1}{3}$, in ascending order of size.
2. Use the calculator to convert the common fractions to decimal fractions (2 d.p.) and arrange in ascending order. Compare the results.

Equal fractions

Finding equal fractions is a **general** method for comparing common fractions. We have already seen that multiplying by $\frac{10}{10}, \frac{100}{100}$, cannot alter a number because we are in effect multiplying by 1. We can extend this to multiplying a number by $\frac{2}{2}, \frac{7}{7}, \frac{5}{5}$ and so on. This 'operation' is very useful for working with common fractions to produce 'equal' fractions.

EXAMPLES

Write $\frac{1}{2}$ as several equal fractions.

$$\frac{1}{2}\left(\times\frac{5}{5}\right)=\frac{5}{10}; \quad \frac{1}{2}\left(\times\frac{4}{4}\right)=\frac{4}{8};$$

$$\frac{1}{2}\left(\times\frac{3}{3}\right)=\frac{3}{6}; \quad \frac{1}{2}\left(\times\frac{6}{6}\right)=\frac{6}{12}; \text{ and so on.}$$

Do you see the idea? We can also reverse this process and divide in effect by 1, i.e.

$$\frac{6}{12}\left(\div\frac{6}{6}\right)=\frac{1}{2}, \quad \frac{5}{10}\left(\div\frac{5}{5}\right)=\frac{1}{2}.$$

This way of working is called reducing the common fraction to its **lowest terms**.

Equal fractions are effective and simple, and so important for correct addition and subtraction of common fractions.

EXERCISE 2.8 Equal common fractions and lowest terms

1. Write $\frac{1}{5}$ as five equal fractions having a denominator (bottom) of 10, 15, 20, 25, 30, respectively.

2. Write in their lowest terms:

$$\frac{25}{100}, \quad \frac{4}{32}, \quad \frac{5}{25}, \quad \frac{75}{100}, \quad \frac{50}{100}$$

Addition and subtraction

Addition and subtraction of common fractions frequently cause errors. Take care!

We can only add and subtract common fractions if they are 'like' fractions, i.e. they are all halves, or fifths, or tenths and so on, in other words they must have the **same denominator** (bottom). In Chapter 5 on algebra you will find something similar in that we can only add or subtract 'like' terms.

When fractions have the same denominator it is fairly clear using our common sense that $\frac{1}{4}+\frac{1}{4}$ must be $\frac{1}{2}$, for example, and that $\frac{7}{10}-\frac{1}{10}=\frac{6}{10}$.

Now look at some other examples:

EXAMPLES

Some whole numbers have been included in these examples. Simplify the following:

1. $\dfrac{1}{4}+\dfrac{1}{8}+\dfrac{3}{16}$ 2. $\dfrac{61}{64}-\dfrac{3}{16}+\dfrac{3}{4}$

3. $2\dfrac{1}{5}-1\dfrac{2}{13}$

These questions can be done as follows:

$$1.\ \frac{1}{4}+\frac{1}{8}+\frac{3}{16}=\frac{1}{4}\left(\times\frac{4}{4}\right)+\frac{1}{8}\left(\times\frac{2}{2}\right)+\frac{3}{16}$$

$$=\frac{4}{16}+\frac{2}{16}+\frac{3}{16}$$

We have changed each of the fractions on the left-hand side to its equal fraction in $\frac{1}{16}$ths. We can now add, with the result $\frac{9}{16}$.

In practice it is usual to go straight to the common denominator as follows:

$$\frac{1}{4}+\frac{1}{8}+\frac{3}{16}=\frac{4+2+3}{16}$$

$$=\frac{9}{16}$$

You must be sure of what you are doing before you shorten it as above.

$$2.\ \frac{61}{64}-\frac{3}{16}+\frac{3}{4}=\frac{61}{64}-\frac{3}{16}\left(\times\frac{4}{4}\right)+\frac{3}{4}\left(\times\frac{16}{16}\right)$$

$$=\frac{61}{64}-\frac{12}{64}+\frac{48}{64}$$

$$=\frac{97}{64}$$

$$=1\frac{33}{64}$$

$$3.\ 2\frac{1}{5}-1\frac{2}{13}$$

What is the common denominator? If it is not clear after some thought then you can multiply together the individual denominators, i.e. 5×13. The result is bound to be a multiple of each.

$$2\frac{1}{5}-1\frac{2}{13}=2+\frac{1}{5}-1-\frac{2}{13}\quad\text{(writing out in full and taking care with the – sign)}$$

$$=1+\frac{1}{5}\left(\times\frac{13}{13}\right)-\frac{2}{13}\left(\times\frac{5}{5}\right)$$

$$=1+\frac{13}{65}-\frac{10}{65}$$

$$=1\frac{3}{65}$$

Now try the following exercise.

EXERCISE 2.9 Addition of common fractions

Simplify the following giving the answer in the lowest terms where appropriate. Then use the calculator to check in decimal fractions (2 d.p.).

(a) $\dfrac{2}{5}+\dfrac{1}{10}+\dfrac{3}{4}$ (b) $\dfrac{31}{64}-\dfrac{5}{16}+\dfrac{3}{4}$

(c) $2\frac{4}{15} - 1\frac{2}{5} + \frac{2}{3}$ (d) $\frac{8}{9} - \frac{7}{18} - \frac{1}{9}$

(e) $3\frac{5}{8} - 1\frac{1}{2} - 1\frac{1}{4}$

Multiplication and division

Multiplication and division of common fractions are easy to work out in a routine manner. It is important, however, that you understand the process. When we write $2 \times \frac{1}{3}$ we are saying '2 lots **of** $\frac{1}{3}$'. Similarly $\frac{1}{2} \times \frac{3}{4}$ can be thought of as '$\frac{1}{2}$ lot **of** $\frac{3}{4}$'. Figure 2.3 may help you to understand.

Figure 2.3 Multiplying fractions: $\frac{1}{2} \times \frac{3}{4}$

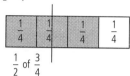

$\frac{1}{2}$ of $\frac{3}{4}$

In the figure three $\frac{1}{4}$'s have been shaded. Do you see that one half of the shaded portion is three eighths of the figure, i.e. $\frac{1}{2} \times \frac{3}{4} = \frac{3}{8}$?

If you repeat this example with different common fractions you will see that an effective rule for multiplication is 'multiply tops, multiply bottoms'. Not a difficult rule to remember!

Look at the following examples.

EXAMPLES

1. $\frac{3}{8} \times \frac{2}{5}$ 2. $\frac{3}{16} \times \frac{2}{9}$ 3. $\frac{1}{13} \times \frac{1}{13}$

1. $\frac{3}{8} \times \frac{2}{5} = \frac{6}{40} \left(\div \frac{2}{2} \right)$

       $= \frac{3}{20}$ (lowest terms)

2. $\frac{3}{16} \times \frac{2}{9} = \frac{6}{144} \left(\div \frac{6}{6} \right)$

       $= \frac{1}{24}$ (lowest terms)

3. $\frac{1}{13} \times \frac{1}{13} = \frac{1}{169}$

You can probably see that the answer in each of the examples is an even smaller common fraction than the one being multiplied. You cannot see this as clearly as you did with the decimal fractions, which are far more easily compared as they are all 'tenths'.

So for both common fractions and decimal fractions we see that **multiplication by a number less than one makes smaller.**

Division is the inverse of multiplication, so that division by $\frac{1}{3}$, say, is the same as multiplication by $\frac{3}{1}$.

EXAMPLES

1. $\dfrac{3}{5} \div \dfrac{1}{5} = \dfrac{3}{5} \times \dfrac{5}{1}$

   $= 3$

2. $\dfrac{3}{5} \div \dfrac{1}{4} = \dfrac{3}{5} \times \dfrac{4}{1}$

   $= \dfrac{12}{5}$

   $= 2\dfrac{2}{5}$

REMEMBER: **division by a number less than one makes bigger.**

When multiplying or dividing with 'mixed' numbers (i.e. including whole numbers) it is best to change to 'top heavy' fractions. For example

$$1\dfrac{1}{2} \times \dfrac{3}{4} = \dfrac{3}{2} \times \dfrac{3}{4} \qquad \text{AND} \qquad 1\dfrac{1}{2} \div \dfrac{3}{4} = \dfrac{3}{2} \times \dfrac{4}{3}$$

$$= \dfrac{9}{8} \quad \text{OR} \quad 1\dfrac{1}{8} \qquad\qquad = 2$$

EXERCISE 2.10 Multiplying and dividing common fractions

Simplify the following:

(a) $\dfrac{3}{8} \times \dfrac{2}{5}$ (b) $\dfrac{5}{6} \times \dfrac{2}{3}$ (c) $\dfrac{3}{4} \times \dfrac{8}{9}$

(d) $\dfrac{3}{8} \div \dfrac{2}{5}$ (e) $\dfrac{5}{6} \div \dfrac{2}{3}$ (f) $\dfrac{3}{4} \div \dfrac{8}{9}$

(g) $2\dfrac{1}{2} \times 1\dfrac{1}{2}$ (h) $1\dfrac{1}{4} \times \dfrac{4}{5}$ (i) $1\dfrac{1}{4} \div \dfrac{4}{5}$

(j) $1 \div \dfrac{1}{6}$ (k) $1 \div \dfrac{3}{4}$ (l) $\dfrac{3}{8} \times \dfrac{2}{5} \div \dfrac{2}{3}$

[Look carefully at your answers to (h), (i), (j) and (k)].

2.6 Checking your calculator answer

You can easily make mistakes when you use your calculator, so it is very important to make a rough estimate of the answer first. For example, in calculating $\dfrac{21.97 \times 0.65}{10.9}$ you should round these mixed numbers to the nearest whole number $\dfrac{22 \times 1}{11} = 2$.

This estimate will not be very accurate, since you have rounded 0.65 to 1. The calculator answer to one decimal place is 1.3, but if you had made a mistake and got 13.1, then your estimate would tell you to have another try!

It is also worthwhile estimating even when you have no decimal fractions, e.g. $\dfrac{7139 \times 23}{117}$ is roughly $\dfrac{7000 \times 20}{100} = 1400$

The calculator answer is 1403.4 to 1 d.p.

Rounded numbers, decimal places and significant figures

There is an important guide rule to rounding off numbers, whether it is just for estimating, or more importantly, for giving an answer to a stated degree of accuracy. The number of decimal places is the number of digits after the decimal point. Take care with zeros.

Suppose a calculator reads 6736.5937. This is cumbersome and we may not need to know it so accurately.

Quoted to 2 **decimal places** (d.p.) it is   6736.59
Quoted to 1 d.p. it is   6736.6
Quoted to the nearest whole number it is 6737
Quoted to the nearest ten   6740
Quoted to the nearest hundred it is   6700
Quoted to the nearest thousand it is   7000

The **rule** is that if the digit dropped is less than 5, then the digit to the left is unchanged, but if the digit is 5 or more, then the digit to the left is increased by one. You may be asked to give an answer rounded to a stated number of decimal places, or in a few cases rounded off to the nearest hundred or thousand.

There is another way in which answers may be asked: that is to a given number of **significant figures**. These are **meaningful** digits. Take care with zeros. Guide rule: this is the number of digits counted from the left to the right, **starting with the first which is not a zero**.

EXAMPLE

All the following have been converted to 3 significant figures (s.f.)
 2076 is   2080 (last zero not significant but essential)
  507.37 is   507
   30.91 is   30.9
    2.887 is   2.89   (note rounding)
    0.87321 is   0.873
    0.06451 is   0.0645
    0.0075571 is   0.00756   (rounding)

EXERCISE 2.11 Estimation and the calculator

1. In each of the following first estimate the answer you would expect and then use the calculator.

(a) $(29 + 31) \times 18$

(b) $(17 + 21) \times 792 \div 11$

(c) $(28 - 12) \times 46 + (39 - 18)$

(d) $(174 - 52) \times 22 \div (61 \times 11)$

(e) $\dfrac{(17 \times 13) - (13 \times 17)}{17 \times 15}$

(f) $(19 + 18) \times 111 \div 3$

(g) $(169 + 30 - 19) \times 23 \div (88 \div 22)$

(h $\dfrac{240 \times 3 \times 17}{340 \times 9}$ (Do you see the quick method?)

2. Using a calculator try multiplication by zero for any numbers. What do you conclude?

3. Using a calculator try division by zero for any number. What do you conclude?

___

Accuracy of measurement

This section is concerned only with checking and correctly giving your answer when using a calculator. The accuracy of a measurement and how to express it is dealt with, together with tolerance, in Chapter 3, Section 3.4.

2.7 Standard form

The decimal place value system provides an easy way to multiply and divide by ten or multiples of ten.

- To multiply a number by 10, (100, 1000, . . .), move the digits of the number 1, (2, 3, . . .) columns to the left.

  For example, 317×10 is 3170

  317×100 is 31 700

- To divide a number by 10, (100, 1000, . . .), move the digits of the number 1, (2, 3, . . .) columns to the right.

  For example, $300 \div 10$ is 30, $300 \div 100$ is 3

NOTE: It is commonly said 'to multiply by ten add a zero', but this is not precise mathematical language, although helpful.

A concise way of writing large numbers is to collect the tens and write them as follows:

$$10 = 10^1 \text{ (we usually omit }^1\text{)}$$
$$100 = 10^2 \text{ means } 10 \times 10$$
$$1000 = 10^3 \text{ means } 10 \times 10 \times 10$$
$$10\,000 = 10^4 \text{ and so on}$$
$$10^6 \text{ is one million}$$

10^1, 10^2, 10^3, . . . are called **powers** of ten and the numbers 1, 2, 3, . . . are called **indices**. So, for example, the index 3 in 10^3 means that 10 is raised to the power 3, and the 3 gives the number of tens multiplied together.

EXAMPLE

The distance of the sun from the earth is approximately 96 million miles. This could be written as 96×10^6 miles. If we were to approximate this further do you see that it is 10^8 miles? (We have 'rounded' 96 to 100.)

For very large numbers (or very small numbers) it is extremely useful to write them as powers of ten in such a way that comparison is easy. This is done by using numbers between 0 and 10 multiplied by the appropriate powers of ten. This way of writing them is called **standard form**.

EXAMPLES

1. Write the following numbers in standard form and then arrange in order of size starting with the smallest.

   (a) 90 000 000   (b) two hundred million

   (c) two hundred thousand

   Written in standard form these are:

   (a) 9×10^7 (b) 2×10^8 ($2 \times 100 \times 10^6$)

   (c) 2×10^5 (200 000)

   In order: 2×10^5, 9×10^7, 2×10^8

2. Write the following in standard form.

   The velocity of light is 330 000 000 m/s

   In standard form this is 3.3×10^8 m/s

EXERCISE 2.12 Standard form

Write the following numbers in standard form and then arrange in order of size, starting with the smallest.
[HINT: Write the numbers in digits first.]

1. Thirty million, three hundred thousand, three hundred million.

2. One hundred and fifty thousand, five hundred thousand, three thousand.

3. 9 000 000 000, 2 000 000, 30 000.

4. The distance from Athens to Barcelona is 3313 km.

5. The earliest marine life and fossils appeared thirteen hundred million years ago.

6. The Sahara has an area of eight million six hundred thousand km².

7. The wavelength of a laser light is 0.000 000 633 m.

2.8 Powers of ten and the operators

Multiplying and dividing

It is very convenient to use powers of ten in calculations involving **multiplication and division** because the powers of ten are easily collected.

EXAMPLES

$10^2 \times 10^3 = 10^5$ (i.e. $10 \times 10 \quad \times \quad 10 \times 10 \times 10$)

$10^1 \times 10^4 = 10^5$ (i.e. $10 \quad \times \quad 10 \times 10 \times 10 \times 10$)

$10^2 \times 10^5 = 10^7$ (i.e. $10 \times 10 \quad \times \quad 10 \times 10 \times 10 \times 10 \times 10$)

You can see that collecting all the 10s means, in effect, that we are **adding the indices when multiplying**.

Now look at division:

$10^3 \div 10^1 = 10^2$ (i.e. $\dfrac{10 \times 10 \times 10}{10}$)

$10^5 \div 10^3 = 10^2$ (i.e. $\dfrac{10 \times 10 \times 10 \times 10 \times 10}{10 \times 10 \times 10}$)

In effect we **subtract the indices when dividing**.

You should now be able to tackle the following exercise. Work quickly but carefully.

EXERCISE 2.13 Powers of ten

Work out the following by writing out the powers in full and then collecting.

(a) $10^3 \times 10^2$ (b) $10^1 \times 10^3$ (c) $10^2 \times 10^2$

(d) $10^3 \div 10^2$ (e) $10^7 \div 10^5$ (f) $10^5 \div 10^2 \times 10^3$

EXERCISE 2.14 Powers of ten

Repeat Exercise 2.13 quickly using the rules for multiplication and division of indices.

You will see in Section 2.11 that the same rules apply for negative indices.

2.9 Powers, roots and reciprocals

Other numbers can also be expressed as powers. For example, 27 can be written as $3 \times 3 \times 3$, i.e. 3^3.

$81 = 9 \times 9 \quad (3 \times 3 \times 3 \times 3)$
$\quad = 3^4$

$4 = 2 \times 2 \qquad 8 = 2 \times 2 \times 2$
$\quad = 2^2 \qquad\qquad = 2^3$

The rules for multiplying and dividing powers of 10 also apply to other numbers raised to powers. For example,

$$3^2 \times 3^3 = 3^5 \qquad 2^3 \times 2^4 = 2^7$$
$$3^5 \div 3^3 = 3^2 \qquad 2 \div 2^4 = 2^{-3}$$

i.e. add the indices for multiplying, subtract the indices for dividing.

In general, using algebraic symbols for any number 'n' and indices 'a' and 'b' we have

$$n^a \times n^b = n^{a+b}$$
$$n^a \div n^b = n^{a-b}$$

This is your first taste of algebra!

Factors, common factors and powers

When a number is written as the product of other numbers these other numbers are called **factors**. For example, $48 = 2 \times 2 \times 2 \times 2 \times 3$ where 3 and the 2s are called factors of 48. Writing 48 in terms of other numbers multiplied together is called **factorizing** 48.

We could also write $48 = 3 \times 2^4$.

Collecting the factors as powers can also help to simplify problems.

EXAMPLES

1. Factorize (a) 121   (b) 196, finding the smallest factors.

2. Simplify $25 \times 10^3 + 75 \times 10^3$ by using a common factor.

   1(a) $121 = 11 \times 11$
       11 has no factors so we cannot factorize further.
       The factors are 11 and 11.

   (b) $196 = 2 \times 98$
       $= 2 \times 2 \times 49$
       $= 2 \times 2 \times 7 \times 7$
       The factors are 2, 2, 7 and 7.
       121 can also be written as 11^2 and 196 can be written as $2^2 \times 7^2$.

   2. $25 \times 10^3 + 75 \times 10^3 = 10^3 (25 + 75)$
                                   $= 10^3 \times 10^2$ (REMEMBER: BODMAS)
                                   $= 10^5$
       10^3 is called a **common** factor.

EXERCISE 2.15 Factorizing

Find the smallest factors of the following and then express as powers. If you know your 'times' tables you should have no problems with this exercise.

(a) 144   (b) 169   (c) 225   (d) 8000   (e) 27 000

(f) 64 000   (g) 1000   (h) 3000   (i) 125   (j) 64

Squares, cubes and roots

The **square** of any number n is the number multiplied by itself, i.e. $n \times n$ or n^2 is the square of n. For example 3×3 or 9 is the square of 3.

The **cube** of any number n is the number multiplied together three times, i.e. $n \times n \times n$ or n^3 is the cube of n. For example $3 \times 3 \times 3$ or 27 is the cube of 3.

The **square root** of any number n is the number which multiplied by itself gives the original number n, i.e. $\sqrt{n} \times \sqrt{n} = n$ where $\sqrt{}$ is the sign for square root. For example $\sqrt{9} \times \sqrt{9} = 9$ where $\sqrt{9}$ is 3.

The **cube root** of any number n is the number which multiplied by itself three times gives the original number n, i.e. $\sqrt[3]{n} \times \sqrt[3]{n} \times \sqrt[3]{n} = n$ where $\sqrt[3]{}$ is the sign for cube root. For example $\sqrt[3]{8} \times \sqrt[3]{8} \times \sqrt[3]{8} = 8$ where $\sqrt[3]{8} = 2$.

Some examples are given in Tables 2.4 and 2.5. These tables form the basis for Exercise 2.16.

Table 2.4 Squares and square roots

Number n	Number n written as square	Square root $\sqrt{n}$
9	3×3 or 3^2	3 or $\sqrt{9}$
16		
25		
36		
100		
144		
169		
225		
400	20×20 or 20^2	20 or $\sqrt{400}$
625		

Table 2.5 Cubes and cube roots

Number n	Number n written as cube	Cube root $\sqrt[3]{n}$
8	$2 \times 2 \times 2$ or 2^3	2 or $\sqrt[3]{8}$
27		
64		
125		
1000		
8000	$20 \times 20 \times 20$ or 20^3	20 or $\sqrt[3]{8000}$

EXERCISE 2.16 Squares, cubes and roots

Copy and complete Tables 2.4 and 2.5.

When you have checked your answers to Exercise 2.16 you will find it helpful to memorize the values.

The rules for multiplying and dividing numbers raised to powers also apply to powers that are not whole numbers,

i.e. $n^a \times n^b = n^{a+b}$ $n^a \div n^b = n^{a-b}$

where a and b can be common or decimal fractions or mixed numbers. For example,

$n^{0.6} \times n^{0.3} = n^{0.9}$ $n^{2\frac{1}{2}} \div n^{1\frac{1}{2}} = n^1$

It is helpful to become familiar with different ways of expressing powers and roots. For example, $n^{3/2}$ can also be written as $n\sqrt{n}$ (i.e. $n^1 \times n^{1/2}$), $\sqrt[3]{n}$ can be written as $n^{1/3}$, $\sqrt[4]{n}$ as $n^{1/4}$, and so on.

Consider the square root of 0.9, i.e. $\sqrt{0.9}$. What do you think it is? This question is a conversation stopper – try it on your friends (or enemies)! The answer that may spring to mind is 0.3? This is **wrong**.

CHECK: 0.3 × 0.3 is 0.09, **not** 0.9.

Think again: $\sqrt{0.81}$ is 0.9 because 0.9 × 0.9 is 0.81. $\sqrt{1.00}$ is 1, therefore $\sqrt{0.9}$ must lie between 0.9 and 1. It must actually be bigger than 0.9, as you should now realize.

Of course you will use the calculator for finding the square roots of mixed numbers, but again it must be emphasized that you need to be able to estimate. So how does one estimate the square root of a decimal fraction such as $\sqrt{0.9}$? Here is a reasonable plan for action:

1. Realize that $\sqrt{0.9}$ will be >0.9 because 'square rooting makes bigger' for numbers less than one.
2. Pair off from decimal point, $\sqrt{0.9} \Rightarrow \sqrt{0.90}$.
3. Think of $\sqrt{90}$, which would be between 9 and 10 (because 90 is between 81 and 100).
4. Put in the decimal point in the right place, i.e. 0.9+.

It is much easier than you may think. It can be done by looking at the number using the critical 'trick' in step 2, i.e. arranging in a pair or pairs from the decimal point.

EXCERCISE 2.17 Estimating square roots

Copy and complete Table 2.6.

Reciprocals

The **reciprocal** of a number n is the inverse of that number, i.e. $\frac{1}{n}$. For example, $\frac{1}{3}$ is the reciprocal of 3, $\frac{1}{7}$ is the reciprocal of 7. Do you see that 3 is the reciprocal of $\frac{1}{3}$, that $\frac{3}{7}$ is the reciprocal $\frac{7}{3}$ and that R is the reciprocal of $\frac{1}{R}$? You will meet the reciprocal in engineering quite frequently, in formulae such as $f = \frac{1}{7}$ (frequency = $\frac{1}{\text{time period}}$), and later for resistors in parallel in an electrical circuit.

REMEMBER: The reciprocal of a number n is $\frac{1}{n}$. This applies not only to whole numbers but also to mixed or fractional numbers. The reciprocal of 0.5 is $\frac{1}{0.5}$,

Table 2.6

Number $n < 1$	$\sqrt{n}$ exact or estimated
0.36	0.6
0.40	between 0.6 and 0.7
0.49	0.7
0.60	
0.70	
0.81	
1.00	
0.94	
0.04	
0.09	
0.12	

i.e. 2, the reciprocal of $\frac{3}{4}$ is $\frac{4}{3}$, and the reciprocal of $\frac{1}{3^{-3}}$ is $\frac{3^{-3}}{1}$, which is $\frac{1}{3^3}$ or $\frac{1}{27}$.

You should be able to work quickly through Exercise 2.18.

EXERCISE 2.18 Finding reciprocals

Write down the reciprocals of the following:

1. 2, 19, 25, n, R

2. $\dfrac{1}{2}$, $\dfrac{1}{3}$, $\dfrac{3}{4}$, $\dfrac{2}{7}$, $\dfrac{1}{n}$

3. $\dfrac{1}{f}$, $\dfrac{I}{V}$, $\dfrac{2}{n}$, $\dfrac{Q}{V}$, ωt, $\sqrt{\dfrac{I}{g}}$

2.10 Integers

Natural or counting numbers are part of a family of **integers**, which can be either positive or negative.

Negative integers were developed to express concisely ideas such as temperatures below freezing point, for example, $-5°C$; depths below a given level, -3 m; financial debt, -- £200; directions plus or minus from a fixed point.

It may help to think of integers on a number line such as that in Fig. 2.4. We use the natural numbers as positive numbers but we do not give them a + sign because we are concerned only with their size.

Figure 2.4 Integers on the number line

Positive and negative integers however indicate 'direction' as well as size and so are called **directed numbers**. There are certain rules of operation which you must therefore learn over and above those already dealt with in this chapter.

Integers and the operators

When operating on positive and negative integers it is advisable to write the integers in brackets in order to avoid confusion with the + and – signs of addition and subtraction.

- When **adding** or **subtracting**, a number line either horizontal or vertical may help. Addition and subtraction can be thought of as movements along the horizontal line or up and down the vertical line.

EXAMPLES

Figure 2.5 Using the number line

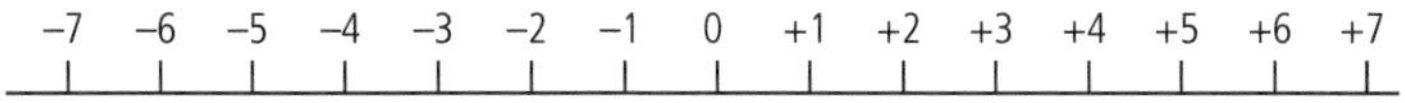

$$-7 \quad -6 \quad -5 \quad -4 \quad -3 \quad -2 \quad -1 \quad 0 \quad +1 \quad +2 \quad +3 \quad +4 \quad +5 \quad +6 \quad +7$$

Refer to Fig. 2.5.

1. $(+2) + (+5) = +7$. (We move to +2 on the number line and then **add** another +5.)

2. $(-2) + (-5) = -7$. (We move to –2 on the number line and then **add** another –5.)

3. $(+5) - (+2) = +3$. (We move to +5 on the number line and then **subtract** +2, i.e. reverse direction.)

4. $(-5) - (-2) = -3$. (We move to –5 on the number line and then subtract –2, i.e. reverse direction.)

Two negatives are not as complicated as they may appear. Other ideas may help, such as **removing** a **debt** of £2, expressed as –(–2), which is the equivalent of £2 credit, expressed as +2. Whatever ideas help you best, remember that –(–) = +. **Subtracting a negative integer is the same as adding a positive integer.**

- When **multiplying** or **dividing** remember that multiplication is repeated addition and division is repeated subtraction. This will help you understand the following examples.

EXAMPLES

1. $(+2) \times (+3) = +6$ (2 lots of (+3))

2. $(-2) \times (-3) = +6$ (two negatives)

3. $(+2) \times (-3) = -6$ (2 lots of (–3))

4. $(+6) \div (+2) = +3$ (how may (+2)s in (+6)?)

5. $(-6) \div (-2) = +3$ (two negatives)

6. $(+6) \div (-2) = -3$ (how may (–2)s in (+6)?)

The BODMAS rule for order of operations applies as it did for the natural numbers. Look carefully at 7.

7. $(+3)[(+5) + (-2)] \div (-3) = (+3) \times (+3) \div (-3)$
$$= (+9) \div (-3)$$
$$= -3$$

For multiplication and division of negative numbers two negatives make a positive. (Look at examples 2 and 5.)

NOTE: to simplify the appearance of a question such as 7, the + sign of a directed integer is usually omitted so that example 7 becomes

$3[5 + (-2)] \div (-3) = 3 \times (3) \div (-3)$
$$= 9 \div (-3)$$
$$= -3$$

EXERCISE 2.19 Directed numbers

Tackle this exercise with confidence.

1. (a) $(+3) + (+2)$ (b) $(+3) + (-7)$
  (c) $(+3) - (+3)$ (d) $(+3) - (+7)$
  (e) $3 - (-7)$ (f) $-3 - (+7)$
  (g) $2 \times (+3)$ (h) $(-2) \times (-3)$
  (i) $7 \times (-3)$ (j) $(-3) \times (+7)$

2. Rewrite in a simpler way and then work out:
$(+5) \times (+4) - [(+7) \times (+2)]$

3. Express the information in the following question as directed numbers and then solve.
[HINT: Start your answer 'morning temp. in °C = ']

In a western Canadian city the temperature on a winter's day was five degrees Celsius below zero. During the night the temperature dropped a further ten degrees Celsius. A wind from the Rockies had by the next morning raised the night temperature by twenty degrees. What was the morning temperature in °C?

2.11 Negative powers and roots

In this section we shall extend the earlier work on powers and roots to negative integers.

Negative powers

Numbers with negative powers are the reciprocals of numbers with positive powers

$$\text{e.g. } 2^{-1} = \frac{1}{2}, \quad 3^{-2} = \frac{1}{3^2}, \quad n^{-a} = \frac{1}{n^a}$$

Negative powers of numbers obey the same rules for multiplication and division as the positive powers.

REMEMBER: In Section 2.9 we saw that

$$n^a \times n^b = n^{a+b}$$
$$n^a \div n^b = n^{a-b}$$

where n is a number (called the 'base') and a and b are indices, i.e. add the indices for multiplication, subtract the indices for division. The following examples include negative values for n, a and b.

EXAMPLES

1. Simplify $2^{-1} \times 2^{-2}$.

   Using the rule, $2^{-1} \times 2^{-2} = 2^{-3}$ (adding indices).

   CHECK that the rule works, knowing that 2^{-1} means $\frac{1}{2}$, and 2^{-2} means $\frac{1}{2^2}$.

   Then $2^{-1} \times 2^{-2} = \frac{1}{2} \times \frac{1}{2 \times 2}$

   $\qquad\qquad\qquad = \frac{1}{2^3}$ OR 2^{-3}

2. Simplify $2^{-1} \div 2^{-2}$.

   Using the rule, $2^{-1} \div 2^{-2} = 2^{-1-(-2)}$ (subtracting indices)

   $\qquad\qquad\qquad = 2^{-1+2}$

   $\qquad\qquad\qquad = 2^1$

   $\qquad\qquad\qquad = 2$

   CHECK: $2^{-1} \div 2^{-2} = \frac{1}{2} \div \frac{1}{2^2}$

   $\qquad\qquad\qquad = \frac{1}{2} \times \frac{2 \times 2}{1}$ (invert and multiply)

   $\qquad\qquad\qquad = 2$

3. Simplify $2^{-2} \times 4^3$.

   $2^{-2} \times (2 \times 2)^3 = 2^{-2} \times 2^3 \times 2^3$ (rearranging to give the same base 2)

   $\qquad\qquad\qquad = 2^4$ (adding indices)

4. Simplify $(-3)^2 \div (-3)^3$.

   $(-3)^2 \div (-3)^3 = (-3)^{-1}$ (subtracting indices)

   $\qquad\qquad$ OR $\quad \frac{1}{(-3)}$ OR $\quad -\frac{1}{3}$

REMEMBER: You can only add and subtract indices when the **same** number (base) is raised to a power.

Negative roots

We saw in Table 2.4 a list of square roots of the natural numbers, i.e. $\sqrt{9}$ is 3, $\sqrt{16}$ is 4. If you think carefully you will now realize that, if we drew up a table with positive integers, there would be two roots equal in size but one would be negative and the other positive. Think of the integer +4. The square roots could be +2 and –2. Why? Because $(-2) \times (-2)$ is +4.

There are many instances in engineering where we are dealing with quantities which have both size and direction and the negative square root has to be included as a solution. There are other situations in which a negative square root would not be an admissible solution, for example because of a particular practical context, and so would be discarded from the final answer.

EXERCISE 2.20 Square roots and indices

1. Copy and complete the following table!

Table 2.7 Positive and negative square roots of some positive integers

Integer	+ve $\sqrt{}$	−ve $\sqrt{}$
+9	+3	−3
+16		
+25		
+49		
+64		
+81		
+100		
+400		
+625		
+900		

Note that all the integers in this table in the first column are positive. Finding the square roots of negative integers is beyond the scope of this book. It takes us into yet another kind of number, imaginary numbers, which, yes, do have applications in electronics!

2. Simplify the following using the rules for addition and subtraction of indices when multiplying or dividing numbers raised to powers.

NOTE: The + sign can be omitted from positive indices.

(a) $3^{-1} \times 3^{-3}$ (b) $5^{-2} \times 5^{+12}$

(c) $2^{-3} \times 4^{-1}$ (d) $3^{-2} \times 9^{+2}$

(e) $3^{-1} \div 3^{-2}$ (f) $5^{-2} \div 5^7$

(g) $10^{-1} \times 10^{-6}$ (h) $10^{-3} \div 10^{-9}$

(i) $(-2)^2 \times (-2)$ (j) $(-3)^2 \div (-3)^{-1}$

Chapter review

Read carefully through the Summary, then test your understanding by trying Exercises 2.21 and 2.22.

Summary

1. The number system in common use is the **decimal** system with digits 0 to 9. All numbers can be expressed as combinations of digits.

2. A fraction is a 'part of a whole'. **Common fractions** are used in everyday language such as one-half ($\frac{1}{2}$), three quarters ($\frac{3}{4}$). The value of a fraction does not change if the top and bottom are divided or multiplied by the same number, e.g. $\frac{1}{2} \times \frac{5}{5} = \frac{5}{10}, \frac{5}{10} \div \frac{5}{5} = \frac{1}{2}$.

3. **Decimal fractions** are common fractions in tenths, e.g. $0.5 = \frac{5}{10}$, 0.75 is $\frac{75}{100}$. Decimal fractions form part of the place value system.

4. BODMAS (brackets, $\div \times + -$) gives the **order** of operations.

5. **Estimate** before using a calculator. Think what happens to fractions when using the $+, -, \times, \div, \sqrt{}$ operators. There are surprises: **multiplication** by $n < 1$ causes **decrease**, division by $n < 1$ causes **increase**, e.g. $9 \times 0.1 = 0.9$, $9 \div 0.1 = 90$.

6. Very small numbers and very large numbers are best expressed in **standard form** as ($n \times$ a power of ten) where $1 \le n < 10$, e.g. $3\,000\,000 = 3 \times 10^6$, $0.000\,003 = 3 \times 10^{-6}$. Engineers use 10^3, 10^6 etc. for large numbers and 10^{-3}, 10^{-6} etc. for small numbers, e.g. 25×10^6 pascals.

7. When a number n is raised to a power, rules for multiplying and dividing are

$$n^a \times n^b = n^{a+b} \quad n^a \div n^b = n^{a-b} \quad n \times n = n^2 \quad \sqrt{n} \times \sqrt{n} = n.$$

8. Integers are whole numbers which can be positive or negative (**directed**).

EXERCISE 2.21 A mixed collection

1. The maximum distance of the planet Mercury from the sun is approximately seventy million km. Write this as a number (i.e. with digits) and then express it in standard form.

2. Substituting into a formula for electrical energy gives the following result:

$$\frac{27 \times 9 \times 9 \times 103}{10^2}$$

First estimate the value and then use your calculator to find the accurate value.

3. A ball is at a height of 20 m at a time t. If $5t^2$ is + 20 what is t? Discuss your solution.

4. Express the following situation in terms of directed numbers. Do not work it out. The temperature in Calgary was 0°C at 8.00 hours, rose 3° by noon, rose another 2° by 15.00 hours then dropped 15° during the night.

5. Express a pressure of seventy thousand million Pa as a number multiplied by the ninth power of ten.

6. Voltage in volts = current in ampères × resistance in ohms. Calculate the resistance of a conductor through which a current of 0.16 ampères is driven by 8 volts.

7. What does the digit 3 represent in the number 1017.039?

8. (a) Work out $\dfrac{0.35 \times 2.4}{0.07}$ without a calculator.

[HINT: Multiply top and bottom by 100.]

(b) Now check your answer using a calculator.

9. A student calculates $\dfrac{7.23 \times 59.6}{17}$ as 253.48.

    (i) Estimate the answer to see whether it is about right or obviously incorrect.

    (ii) Use a calculator to give the result to 1 d.p.

10. Express the following in standard form:

    (a) 310 000; (b) $18.5 \times 100\,000$; (c) 0.0033×1000; (d) 0.0013; (e) 0.000 000 392

11. The distance of the planet Pluto from the sun is 7364 million kilometres. Express the distance in standard form.

12. Express the following decimal fractions as common fractions in lowest terms:

    0.02, 0.75, 0.125, 0.375, 0.675

13. Without a calculator express the following common fractions as decimal fractions:

    [HINT: Multiply or divide top and bottom by the same number, e.g. $\dfrac{35}{50} \div \dfrac{5}{5} = \dfrac{7}{10}$ i.e. 0.7.]

$$\frac{1}{5}, \frac{2}{5}, \frac{3}{5}, \frac{4}{5}, \frac{1}{25}, \frac{3}{25}, \frac{5}{25}, \frac{10}{25}, \frac{15}{50}, \frac{25}{50}$$

14. Without a calculator express the following mixed fractions as mixed decimals.

    [HINT: Change only the common fraction part, e.g. $1\dfrac{4}{5}$ is $1 + \dfrac{4}{5} \times \dfrac{2}{2} = 1 + \dfrac{8}{10} = 1.8$]

$$1\frac{1}{2}, \ 1\frac{1}{4}, \ 2\frac{3}{5}, \ 3\frac{3}{4}, \ 5\frac{1}{25}, \ 1\frac{3}{25}$$

15. Without a calculator work out the following and then arrange the answers in ascending order of size.

    (a) 0.125×0.5 (b) 0.5×0.25 (c) 0.25×0.25 (d) $\sqrt{0.0625}$ (e) $\sqrt{0.09}$

16. Use your mental agility with number to work out the following.

    [HINT: Multiply top and bottom by the same number if it helps you, e.g. 10 or 100]

$$\frac{10}{0.5}, \ \frac{90}{0.3}, \ \frac{100}{0.25}, \ \frac{75}{0.75}, \ \frac{1.25 \times 100}{0.5}.$$

The following exercise will give you practice in calculating with numbers that could arise from measurements and given data.

EXERCISE 2.22 Using a calculator

In each of the following, first estimate the answer you would expect. [HINT: You may need to divide top and bottom by the same number, i.e. the 'cancelling' technique]. Then use the calculator and give the answer to 2 d.p. where appropriate.

Take $\pi = 3.142$

1. $2\pi\sqrt{\dfrac{0.52}{9.81}}$ 2. $\sqrt{\dfrac{441}{\pi}}$

3. $\dfrac{410}{2\pi}$ 4. $240(15.3 + 17.5) - 240 \times 12.5$

5. $\pi \times (10.3)^2 \times 15.1$

6. $\dfrac{14.5 \times 5.5}{(27.3 + 18.7)}$

7. $\dfrac{1}{3} \times (6.5)^2 \times 9.7$

8. $\dfrac{\pi}{6}(5.7)^3$

9. $4.7 \times 2.1 \times 1.5$

10. $\dfrac{628 \times 3.59}{\pi(10.2)^2}$

chapter

3

Ratio, proportion and percentage

3.1 Ratio

Ratio is the relative measure of two quantities. If the quantities are a and b we write the ratio as $a:b$ (spoken as 'a to b').

For example, if there are 7 female students and 14 male students on a course then:

1. the ratio of females to males is $7:14$ or, more simply, $1:2$.
2. the ratio of males to females is $14:7$ or $2:1$.
3. the ratio of females to total students is $7:21$ or $1:3$.

Since ratio is a relative measure and therefore expressed only in numbers, you must take great care with the units when comparing quantities. For example, if you wish to compare a distance of 1 metre with a distance of 1 kilometre, you **cannot** write it as $1:1$. The distances must be in the **same units** for comparison. If the unit chosen is the metre then the ratio is $1:1000$. If the unit chosen is the kilometre then the ratio is $\frac{1}{1000}:1$, which is the same as $1:1000$. You will find ratio in scale drawings, maps and models when it is usual to write ratio in the form $1:n$.

Ratio is a straightforward idea yet there are pitfalls in apparently simple tasks.

EXAMPLE

1. On a plan where 10 mm represents 1 m, what is the scale ratio, i.e. the ratio, $\dfrac{\text{length on plan}}{\text{true length}}$?

To compare a plan length with the true length the units must be the same. Let us choose mm. Then, scale ratio is 10 mm : 1000 mm which is 1 : 100.

When a ratio is the comparison of a part to the **whole** it has the same meaning as a fraction.

EXAMPLES

1. If the ratio of female students to total number of students is 1 : 5 this is the same as saying that the female fraction of the total is $\frac{1}{5}$.

Note the everyday use of the common fraction. It would appear strange to use the decimal fraction, 0.2 in this case.

2. If the ratio used in building a model railway is 1 : 100, then a length on the model will be $\frac{1}{100}$ of the true length.

In practice, for calculations, all ratios may be regarded as fractions.

EXERCISE 3.1 Ratio

1. A mixture is composed of seven parts of liquid A and six parts of liquid B by volume. Find the ratio by volume of:
   (a) liquid B to liquid A
   (b) liquid A to the total volume
   (c) liquid B to the total volume.

2. In an architectural plan the scale used was 1 : 50. What would be the true length in metres corresponding to a length of 10 mm on the plan?

3. On an Ordnance Survey map a distance of 500 m is represented by 10 mm. What is the scale ratio?

4. The population of a village A is 2000 whilst that of a nearby town B is 400 000. What is the ratio of the population of A to B?
   (Write your answer in the form 1 : n)

5. The ratio of the resistances of two electrical conductors A and B is 1 : 3. If the resistance of B is 6 ohms, what is the resistance of A?

6. An alloy consists of 27 g of copper and 3 g of tin. Find the ratio by mass of:
   (a) copper to tin
   (b) copper to the alloy

While in everyday life we usually deal with ratios of like quantities, for example, length on map : length on ground, in engineering we meet constant ratios between unlike quantities. For example, for a material under fixed conditions the ratio mass : volume is constant. The quantity mass/volume is the density of the material. You will meet other examples.

3.2 Proportion and ratio

If two variable quantities are in **direct proportion** the ratio between them is always the same (i.e. is constant).

If the quantities are a and b, then $a \propto b$ means

$$a : b \text{ or } \frac{a}{b} = \text{constant}$$

$$\Rightarrow \frac{a_1}{b_1} = \frac{a_2}{b_2} \quad \text{OR} \quad \frac{a_1}{a_2} = \frac{b_1}{b_2}$$

This means that 'if a is doubled, b is doubled', 'if a is halved, b is halved'.

When a vehicle travels at constant speed, for example, the distance covered,

d, is directly proportional to time, t, i.e. $d \propto t$. This means that three times the distance is covered in three times the time, and so on.

EXAMPLES

1. This example shows you how to calculate an unknown value in a problem involving proportion by two different methods.

   A train travelling at constant speed covers 180 miles in 3 hours. How far has it travelled in 2 hours?

   $$\frac{distance_1}{time_1} = \frac{distance_2}{time_2} \quad \text{(i.e. speed stays the same)}$$

   $$\frac{180}{3} = \frac{d}{2} \quad \text{where } d \text{ is the distance covered in 2 hours.}$$

   $$60 = \frac{d}{2}$$

   $$d = 120 \text{ miles}$$

   OR using the second method (called the unitary method because we find 1 unit in the middle of the calculation):

   In 3 hours a train covers 180 miles

   In 1 hour a train covers $\frac{180}{3}$ i.e. 60 miles

   In 2 hours a train covers 60×2 i.e. 120 miles

   [HINT: Put the unknown, i.e. distance, at the end of the statement.]

2. The extension of a wire obeying Hooke's Law is 1.73 mm when the load is 12.0 N. Calculate the extension when the load is 37.0 N.

   $$F_1 = 12.0 \text{ N} \quad x_1 = 1.73 \text{ mm}$$
   $$F_2 = 37.0 \text{ N} \quad x_2 = ?$$

   Hooke's Law states $x \propto F$

   $$\Rightarrow \frac{x_2}{x_1} = \frac{F_2}{F_1} \quad \text{[Keep the unknown on the top]}$$

   $$\frac{x_2}{1.73} = \frac{37}{12}$$

   $$x_2 = \frac{1.73 \times 37}{12}$$

   $$= 5.33 \text{ mm}$$

   The extension is 5.33 mm

EXERCISE 3.2a Direct proportion

1. A car travelling on the motorway covers a distance of 140 miles in 2 hours. Assuming that it is travelling with constant speed, find how far it will have travelled in 3 hours.

2. A mixture is composed of seven parts of liquid A to six parts of liquid B by volume. In 390 mL of the liquid what is the volume occupied by each liquid?

3. For a given mass, force F is directly proportional to the acceleration a, i.e. $F \propto a$. If a force of 16.0 N causes an acceleration of 20 m/s², what force will give an acceleration of 30 m/s²?

4. A and B invest money in a small business. A put in £5000 and B put in £3000. The profits are divided in the same ratio as their investments. The profit at the end of the first year was £12 000. How much did each receive?

5. The ratio of the sizes of two spanners is 4:7. If the larger is 14 mm what is the size of the smaller?

Inverse proportion

If two variable quantities are in **inverse proportion**, then an **increase** in one produces a proportional **decrease** in the other.

If the quantities are a and b, then $a \propto \dfrac{1}{b}$ means $a \times b$, i.e. ab, is constant

$$\Rightarrow a_1 b_1 = a_2 b_2 \quad \text{OR} \quad \frac{a_1}{a_2} = \frac{b_2}{b_1}$$

This means that 'if a is doubled, b is halved', and 'if a is halved, b is doubled'.

Some examples are:
- When travelling a given distance, the greater the speed u, the less the time taken t, where ut is constant.
- For a vibrating spring the greater the length l, the lower the frequency f, where fl is constant.
- For a given voltage the greater the resistance R, the smaller the current I, where RI is constant.

EXAMPLES

1. A mass of 5 kg is accelerated at 2 m/s² by a force F. What is the acceleration when the same force F acts on a mass of 10 kg?

$a \propto \dfrac{1}{m}$ if F is constant, i.e. ma is constant

Thus, if m is doubled, a is halved

$\Rightarrow$ acceleration of 10 kg $= \dfrac{2 \, m/s^2}{2}$

$= 1 \, m/s^2$

2. The resistance R of a piece of wire is 8.0 Ω and its cross-sectional area a is 5.0 mm². What is the cross-sectional area if $R = 13.0$ Ω? Assume the length l and the material are the same.

$R \propto \dfrac{1}{a}$ if l is constant

$\Rightarrow R_1 a_1 = R_2 a_2$ where $R_1 = 8.0$ Ω $\quad a_1 = 5.0$ mm²
and $R_2 = 13.0$ Ω $\quad a_2 = ?$

$$8 \times 5 = 13 \times a_2$$

$$a_2 = \frac{8 \times 5}{13}$$

$$= 3.1 \text{ mm}^2 \quad (2 \text{ s.f.})$$

EXERCISE 3.2b Inverse proportion

1. If a is inversely proportional to b and a is 7.5 when b is 6, what is a when b is 3?

2. A gas obeying Boyle's law (pressure × volume = constant) is compressed from a volume of 0.060 m³ as the pressure is increased from 1.4×10^6 Pa to 4.5×10^6 Pa, the temperature remaining constant. Calculate the new volume.

In Chapters 7 and 8 you will see that relationships between variable quantities can also be expressed through the use of graphs. In Chapter 7 you will see direct proportionality and inverse proportionality illustrated by graphs.

3.3 Percentage and ratio

There is scarcely any aspect of living or work where percentage does not play a part. And yet many people, including students, find difficulty in understanding and using percentage correctly.

A percentage can be thought of as a special ratio where the total number of parts is 100. For example, 20 per cent, written concisely as 20%, **means 20 for each 100**. This can also be written as $\frac{20}{100}$.

Thus 20% of 100 is 20
20% of 200 is 40, and so on.

To calculate 20% of any number we multiply the number by $\frac{20}{100}$.

EXAMPLES

Find 1. 20% of £1000   2. 20% of £2000
3. 25% of £1000   4. 25% of £2000

In calculating it helps to replace 'of' by 'multiply'.

1. $20\% \times £1000 = £\frac{20}{100} \times 1000$
$= £200$

2. $20\% \times £2000 = £\frac{20}{100} \times 2000$
$= £400$

3. $25\% \times £1000 = £\frac{25}{100} \times 1000$
$= £250$

4. $25\% \times £2000 = £\dfrac{25}{100} \times 2000$

$= £500$

If you look at these examples carefully you will see that the **same percentage** can represent quite **different** sums of money. It is essential to ask 'percentage of WHAT?'

EXERCISE 3.3 Finding percentages

Do not use a calculator.

1. Find 10% of (a) £1000 (b) £15 000

2. Find 70% of (a) 100 marks (b) 50 marks

3. Find 5% of (a) 100 components (b) 460 components

4. A factory should have 1000 workers present. If 100 are absent what percentage is present?

5. On one jumbo jet scheduled flight, 5% of the 400 seats were unoccupied. Find how many seats were occupied.

Changing common and decimal fractions to percentages

We have already said that a percentage was a special kind of ratio i.e. so many parts per 100. We can also think of percentage as a special kind of fraction where the total number of parts forming the whole is 100. Thus 30% is $\frac{30}{100}$, which in its lowest terms is the common fraction $\frac{3}{10}$. Note that if you have £100 and spend £30 you have spent 30% of your money.

To change a ratio or a common fraction to a percentage, multiply by 100. To change a decimal fraction to a percentage we also multiply by 100. For example, 0.175 means $\frac{175}{1000}$ or $\frac{17.5}{100}$, i.e. 17.5% (17.5 parts in 100). In effect, we have multiplied 0.175 by 100. The general rule therefore applies to both common and decimal fractions, i.e. **multiply by 100 to convert to a percentage**.

EXAMPLES

1. Change 0.475 to a percentage.
2. Change $\dfrac{3}{40}$ to a percentage.

  1. $0.475 = 0.475 \times 100\%$

       $= 47.5\%$

  2. $\dfrac{3}{40} = \dfrac{3}{40} \times 100\%$

       $= \dfrac{3}{4} \times 10\%$ (dividing top and bottom by 10)

       $= \dfrac{3}{2} \times 5\%$ (dividing top and bottom by 2)

       $= 7\dfrac{1}{2}\%$

NOTE: In question 2:

(a) You should be able to divide top and bottom by 20 and omit the middle line.
(b) If the numbers are awkward use a calculator but do not use the calculator % button.

EXERCISE 3.4 Changing fractions to percentages

Do not use a calculator.
Convert the following to percentages:

1. 0.25,   0.80,   1

2. 0.325,   0.675,   0.932

3. $\dfrac{1}{10}$, $\dfrac{1}{25}$, $\dfrac{1}{40}$

4. $\dfrac{1}{2}$, $\dfrac{1}{4}$, $\dfrac{1}{8}$

Changing percentages to common or decimal fractions

This is the reverse of the process of changing to percentages.

EXAMPLE

Express $17\frac{1}{2}\%$ as a common fraction and as a decimal fraction.

As a common fraction:

$$17\frac{1}{2}\% = \frac{17\frac{1}{2}}{100} \text{ (i.e. } 17\frac{1}{2} \text{ parts per 100)}$$

$$= \frac{35}{200} \left(\div \frac{5}{5} \right)$$

$$\text{OR} \quad \frac{7}{40} \text{ (in lowest terms)}$$

As a decimal fraction:

$$17\frac{1}{2}\% = \frac{17.5}{100}$$

$$= 0.175$$

Thus we have three ways of expressing the same number

i.e. $17\frac{1}{2}\%$ OR $\dfrac{7}{40}$ OR   0.175

NOTE: It is worth memorizing these values because the current rate of VAT is $17\frac{1}{2}\%$.

You will now understand that VAT, interest rates etc. are quoted as percentages because of the easy comparison.

EXERCISE 3.5 Changing percentages to fractions

Do not use a calculator.
Express each of the following both as a decimal fraction and a common fraction in its lowest terms.

1. (a) 10%   (b) 5%   (c) 8%   (d) 25%
2. (a) 7.5%   (b) 57.5%   (c) 33%

Table 3.1 Decimal and common fractions corresponding to some percentages

%	5	10	20	25	50	75	100	$33\frac{1}{3}$	$66\frac{2}{3}$	$17\frac{1}{2}$
Decimal fraction	0.05	0.1	0.2	0.25	0.5	0.75	1	$0.\dot{3}$	$0.\dot{6}$	0.175
Common fraction	$\frac{1}{20}$	$\frac{1}{10}$	$\frac{1}{5}$	$\frac{1}{4}$	$\frac{1}{2}$	$\frac{3}{4}$	$\frac{1}{1}$	$\frac{1}{3}$	$\frac{2}{3}$	$\frac{7}{40}$

Table 3.1 gives you some values which are worth memorizing. You will notice that the table includes $\frac{1}{3}$ and $\frac{2}{3}$ written as $0.\dot{3}$ and $0.\dot{6}$. If you use the calculator (or use your head) to convert $\frac{1}{3}$ and $\frac{2}{3}$ into decimal fractions you will obtain 0.33333 ... and 0.66666 ... . There is no end to the digits and they are called **recurring decimals** written in the special way as $0.\dot{3}$ and $0.\dot{6}$.

More surprises with percentages

First surprise: greater than 100%
Do you realize that it is possible to have a percentage greater than 100%? If, for example, we convert the mixed number 1.43 to a percentage it becomes 143% (1.43×100)!

If we convert 2 it becomes 200%! Look at the following example **carefully**.

EXAMPLE

Figure 3.1 Product sales for consecutive years

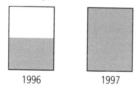

1996   1997

Figure 3.1 shows the sales of a product for two successive years, 1996 and 1997.
(a) What is the new percentage compared with the original?
(b) What is the percentage increase in the second year?

(a) The sales have doubled and so the **new percentage** is 200%.
(b) If the sales for the first year are 100% and for the second year 200% the **percentage increase** is 100%.

Second surprise: discounted percentages
Percentage discounts deserve special mention since they appear to cause much confusion and therefore error in the business and technological fields or in any application where a % discount operates as in VAT calculations. The secret is always to represent the original price as 100%.

EXAMPLE

An engineering invoice showed £434.75 including VAT at $17\frac{1}{2}$%. Find the price before VAT was added.

Look carefully at the solution: it involves proportion.

117.5% is £434.75

100% is £434.75 × $\dfrac{100}{117.5}$

Using a calculator: price is £370

Look at the answer. Does it make sense? Yes, because we expect 100% to be a smaller sum of money than 117.5%. Common sense still does not tell us whether our solution is correct so we can check it as follows.

CHECK: 100% = £370

17.5% = £ $\dfrac{17.5}{100}$ × 370

⇒ VAT = £64.75
⇒ Price +VAT = £370 + £64.75
= £434.75

We shall do one more example involving cost price and selling price before you tackle the next exercise. In problems involving profit, loss and discount, it is usual to call the cost price 100%.

EXAMPLE

A bicycle when new costs £234.50 including VAT. Two years later it is sold for £65. Find the percentage loss. Use a calculator.

Cost price = £234.50   100%
Selling price = £65   ?
⇒ Loss = £169.50

% Loss = $\dfrac{169.5}{234.5}$ × 100 (fraction of cost × 100)

% Loss = 72.3 (1 d.p.)

Using a calculator to find percentages

If you had pressed a calculator button for the exercises on percentage you would not now be understanding percentages and how to use them. When you use a calculator it is better to **think** what a percentage means. So if you were finding 20% of 50, say, proceed as follows: 20 ÷ 100 × 50 = 10.

If you were finding $\frac{1}{5}$ as a percentage, you would proceed as follows:

1 ÷ 5 × 100 = 20%.

When, and only when, you have 'mastered' percentages should you think of using the calculator button.

EXERCISE 3.6 Mixed percentages

Use your calculator where needed.
1. Two people, one earning £80 a week and the other £250 a week are each awarded a 5% rise. What is each new **increase** in wage?

2. An aircraft flew 10% faster than the speed of sound (750 mph). What was the speed of the aircraft?

3. In a test my ratio of correct answers to incorrect answers was 3:1. (a) How many correct answers did I get if the test had 40 questions? (b) What percentage of my answers was correct?

4. If the sales of a product treble in two successive years, find:
   (a) the new percentage of the original
   (b) the percentage increase. First draw a sketch.

5. A laptop computer with VAT included at 17.5% is priced at £1527.50. Find:
   (a) the cost before VAT is added
   (b) the amount of VAT.

6. In a sample of 55 components, three were found to be defective. Find the percentage of defective components (2 d.p.).

7. A flat was bought for £39 475 and three years later sold for £51 950. Find the percentage profit (2 d.p.).

8. An alloy consists of 55% copper, 28% zinc and 17% nickel. Find the masses of copper, zinc and nickel in a 3.5 kg block of the alloy.

9. Express the following percentages as common fractions in their lowest terms:
   $12\frac{1}{2}$%, 30%, 85%.

10. Express the following percentages as decimal fractions:
    $37\frac{1}{2}$%, 1%, $\frac{1}{2}$%.

11. Convert the following test scores to percentages and hence find the best score:
    $\frac{17}{30}$, $\frac{21}{50}$, $\frac{42}{75}$

12. Express the following as percentages:
    0.075, 1.5, $\frac{4}{15}$

3.4 Accuracy and tolerance

In the last chapter we were concerned with checking your calculator answer against human error and with giving an answer to a stated degree of accuracy expressed by decimal places or significant figures. You learnt also how to round off numbers as required.

In this section we shall consider the accuracy of measurements, and when these are used in a calculation, **knowing** the level of accuracy to which your answer can be given.

Accuracy

If you measure a rectangle with a ruler which measures to the nearest millimetre, you might obtain sides of 37.6 cm and 6.4 cm. The area by calculator is

$$37.6 \times 6.4 = 240.64 \text{ cm}^2$$

But suppose you had an instrument accurate to $\frac{1}{10}$ mm. For 37.6 cm you could find any measurement between 37.55 cm and 37.64 cm. For 6.4 cm a value between 6.35 cm and 6.44 cm. These numbers are in the range which your ruler would round off to 37.6 cm and 6.4 cm. Using these upper and lower limits your calculator would give:

(a) $37.64 \times 6.44 = 242.4016 \text{ cm}^2$
(b) $37.55 \times 6.35 = 238.4425 \text{ cm}^2$

To the nearest whole number the true area lies between 242 cm^2 and 238 cm^2, so it would be quite wrong to give your answer from the ruler measurements as 240.64 cm^2 – the ruler is not that accurate!

How should the answer be expressed? Now 37.6 is to 3 s.f. and measured to the nearest mm, and so is correct to 0.1 in 37.6, i.e. 1 in 376. Also 6.4 is to 2 s.f. and is correct to 0.1 in 6.4, i.e. 1 in 64. The smaller measurement is clearly less accurate.

The area, between 242 cm^2 and 238 cm^2, should be rounded to 240 cm^2, which is to **2** s.f.

The guide rule when multiplying or dividing is that your answer should be correct to the **same number of significant figures as the least accurate measurement**.

Adding or subtracting measurements is a matter of simple common sense. If an instrument is correct to 0.1 mm, then a series of measurements added or subtracted are also correct to 0.1 mm

e.g. $10.67 \text{ cm} + 41.03 \text{ cm} + 6.32 \text{ cm} = 58.02 \text{ cm}$.

But if you had measured the first of these numbers with a ruler, correct to only 1 mm, you would have

$10.7 \text{ cm} + 41.03 \text{ cm} + 6.32 \text{ cm} = 58.05 \text{ cm}$.

This should be quoted as 58.1 cm, since the lower accuracy of the first measurement has reduced the accuracy of the total.

EXERCISE 3.7 Significant figures and decimal places

Work out the following, giving your answers to the appropriate number of s.f. or d.p.
1. $14.2 \times 3.65 \times 1.25$
2. $0.019 \times 101 \times \pi$
3. $4\pi r^2$, where $r = 0.15 \text{ m}$
4. $7.2 \times 0.013 \div 150$
5. $32.7 - 1.64$

Methods of stating accuracy of measurements

Scientific measurements are usually expressed as follows:

e.g. temperature $76°C \pm 1°C$ or $76 \pm 1°C$

This means the value lies between 77°C and 75°C.

Any scientific reading from which conclusions are derived must be measured with a known degree of accuracy, otherwise it is of little value.

It is also possible to express accuracy as a **percentage** (see Section 3.3 for percentages). If a length of 5 mm is measured with an accuracy of 1 mm, this can be expressed as 5 ± 1 mm. But ± 1 mm in 5 mm is an error of 1 in 5, which is the same as 20 in 100, i.e. 20%. A very poor accuracy! However, the same error in 100 mm measured is

$$100 \pm 1 \text{ mm} \quad \text{or} \quad 1 \text{ in } 100 \quad \text{or} \quad 1\% \text{ error}$$

To achieve good accuracy in a small measurement, you need a very accurate measuring instrument.

EXAMPLE

A pressure gauge has an accuracy of ± 0.05 bar. What is the percentage error when it reads 4.6 bar?

Reading is expressed as 4.6 ± 0.05 bar

$$\% \text{ error} = \frac{0.05}{4.6} \times 100$$
$$= 1.1\% \text{ (2 s.f.)}$$

Tolerance

When components are manufactured, it is likely that there will be a stated value for resistance, diameter or any other measurement in the specification. In practice, not all will match exactly the nominal value but there is a stated range about that value which allows the product to be used. Samples falling outside that range would be rejected.

The allowed range around the preferred value is called the tolerance. It can be expressed as a physical measurement or as a percentage.

EXAMPLES

1. The given value of a resistor is $680 \text{ k}\Omega \pm 5\%$. Does a resistor of $642 \text{ k}\Omega$ meet the specification?

$$5\% \text{ of } 680 \text{ kW} = \frac{5}{100} \times 680 \text{ k}\Omega$$
$$= 34.0 \text{ k}\Omega$$

Lower limit of tolerance $= (680 - 34) \text{ k}\Omega$
$$= 646 \text{ k}\Omega$$
$642 \text{ k}\Omega$ is below this lower limit

The resistor does NOT meet the specification.

What is the upper limit of tolerance?

Upper limit $= (680 + 34) \text{ k}\Omega$
$$= 714 \text{ k}\Omega$$

2. The diameter of a hole in a working part is $16 \text{ mm} \pm 0.05$ mm. What is the smallest acceptable diameter?

Smallest diameter $= 16 \text{ mm} - 0.05 \text{ mm} = 15.95 \text{ mm}$.

Tolerance can be specified for any measurable quantity, for example mass, time, current, hardness, diameter, composition, temperature, etc. Controlling items to be within a suitable tolerance band is very important in manufacturing.

It is no good making a 12 mm shaft to turn in a 12 mm hole. The shaft may work if its diameter is from 11.90 mm up to 11.98 mm and the internal diameter of the hole may be from 12.00 mm up to 12.06 mm. Then the clearance will be between 0.02 mm and 0.16 mm. Thus the specification for the shaft diameter should be 11.94 ± 0.04 mm. Samples are taken during manufacture and there will be a distribution of diameters about the mean value owing to random variation (see Chapter 4). The aim is for 100% of the items to lie within the tolerance band as shown in Fig. 3.2.

Figure 3.2 Variation about the mean within the tolerance band

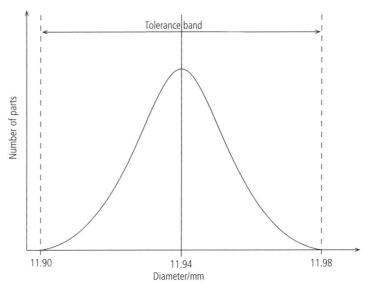

Statistical quality control is comparison of the results of sampling with the theoretical curve. Control charts show warning limits and action limits inside the tolerance band. If a sample mean falls outside the warning limit, the manufacturing process is not in control. How much drift beyond the warning limit is allowed depends on the precision required. Figure 3.3 shows the limits superimposed on the normal distribution and the corresponding control chart with its limits.

EXERCISE 3.8 Tolerance

1. Fixed resistors are colour-coded by coloured bands which indicate the value and the tolerance. The code is shown in Fig. 3.4. What is the value and tolerance of a resistor with bands of
   (a) brown, black, red and silver?
   (b) orange, orange, yellow and red?

Figure 3.3 Control chart superimposed on normal distribution about the mean

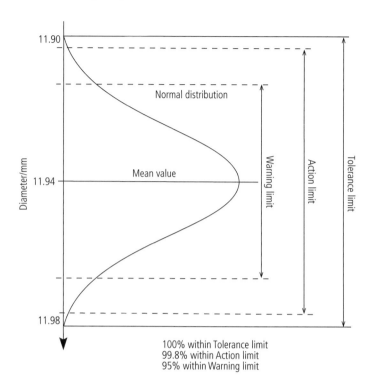

100% within Tolerance limit
99.8% within Action limit
95% within Warning limit

Figure 3.4 Resistor colour codes

Example shown (yellow, violet, orange, gold) = 47kΩ ± 5%

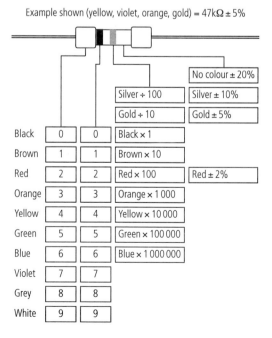

2. You need to use three resistors as follows:
   (a) 5.6 kΩ ± 2% (b) 47 kΩ ± 10% (c) 6.8 MΩ ± 5%
   What coloured bands will each carry?

3. 680 μF capacitors have a tolerance of ± 20%. Calculate the lowest and highest values at the limits of the tolerance band.

4. A diameter of a manufactured part is 10 ± 0.3 mm. What percentage tolerance does this represent? What are the highest and lowest values allowed?

5. The drawing in Fig. 3.5 shows dimensions in mm of a capacitor.
   (a) What is the lowest allowed value for measurement x?
   (b) What is the greatest possible difference between measurement y and z?

Figure 3.5 Tolerances on a capacitor drawing

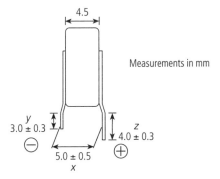

6. The mass of chocolates in a 1 kg box is subject to some variation such that the contents are sometimes up to 5 g over or under specification. What is the percentage tolerance?

Tolerances on drawings

Figure 3.5 shows tolerances on a drawing. Sometimes all dimensions have the same tolerance (of ± 0.02 mm say) and so they are not individually marked. Occasionally, the upper or lower limit of a dimension has no tolerance and the drawing will be marked accordingly, for example + 0.00 mm
                                                                    – 0.02 mm.

Chapter review

Read the summary carefully, checking that you understand what you have learned in this chapter. Then try the final exercise.

Summary

1. Ratio is a relative measure of two quantities, written as $a:b$ or $1:n$ (scale drawings). Ratio can be used as a fraction.

2. (a) Two quantities are in direct proportion if their ratio stays the same

   i.e. if $\dfrac{a}{b}$ is constant then $\dfrac{a_1}{b_1} = \dfrac{a_2}{b_2}$

e.g. $\dfrac{\text{distance}}{\text{time}}$ (constant speed).

(b) Two quantities are in inverse proportion if their product stays the same

i.e. if ab = constant then $\dfrac{a_1}{a_2} = \dfrac{b_2}{b_1}$.

3. Percentage (per 100) is a special ratio or fraction with 100 parts total.

e.g. 20% is 20 in each 100, 20 : 100 or $\dfrac{20}{100}$.

4. Percentages, common fractions and decimal fractions are different forms of number and are **interchangeable**. To change a common or decimal fraction to a percentage multiply by 100, e.g. $\frac{1}{2} = 0.5 = 50\%$.

5. Tolerance is the allowed range round a nominal value, expressed as a physical measurement, such as ±2 mm, or as a percentage, say ±4%.

EXERCISE 3.9 Mixed questions

1. A length of metal expands by 0.000 011 of its original length for each degree Celsius rise in temperature. Calculate the expansion for a temperature rise of five degrees.

2. The ratio of circumference : diameter of a circle is approximately 3.14 : 1. Calculate to the nearest mm the circumference of a circle of diameter 95 mm.

3. A large gear wheel having 20 teeth is connected to a smaller wheel with 10 teeth. What is the ratio of the speed of the large wheel to the speed of the small wheel? (Write in the form 1 : n.)

4. A machine has a force ratio of 40 and a distance ratio of 80.
Efficiency = $\dfrac{\text{force ratio}}{\text{distance ratio}} \times 100\%$. Find the efficiency of the machine.

5. 8 m³ of concrete is to be mixed in the following proportions:
   cement   10%
   sand   45%
   aggregate 45%
   (a) What is the ratio of cement to sand to aggregate? Write your answer as whole numbers taking cement as 2 parts.
   (b) How much aggregate is required?

6. The diameter of a machined shaft is 11.97 mm. The specification is 11.94 ± 0.04 mm. Should this item be rejected? Show your working when you do the calculation.

4

An introduction to statistics

4.1 What is statistics?

Statistics is concerned with the correction, classification and interpretation of information. We are surrounded by data which we need to interpret and use to make decisions. If we are to make sense of the information, it must be presented in a structured way. We employ a variety of methods including tables, charts and graphs.

Interpretation sometimes requires an estimate of the likelihood (i.e. the probability) of something happening. For example, if a survey is conducted one winter of the incidence of flu in the population, and the result is compared with records from previous years, we may conclude that a flu epidemic is likely.

Statistics can mislead. It is very important that you examine information, particularly that presented visually, very closely indeed. Advertisers, for example, are very skilful in making clever, persuasive presentations which are designed to give a favourable impression. You can be misled unless you scrutinize the result carefully.

Choosing a scale which disguises the zero is one way of making results look favourable, as illustrated in Fig. 4.1.

Managerial decisions, market research and quality control are only a few of the activities which rely on statistics. You will at some stage be involved in decision making and your care in interpretation may make the difference between profit and loss.

4.2 Statistical terms

Data

Data are pieces of information. They may be given to you or you may collect them yourself.

Figure 4.1   A 'profit surge' which is no more than a seasonal fluctuation

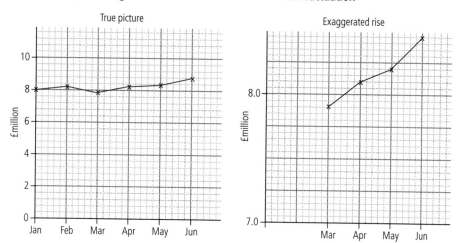

NOTE: A 'data base' or 'data bank' is a large store of information usually in a form for a computer to handle (in which case a 'database').

Variable

A variable is something which can change, i.e. it has a range of possible values. Some examples are height of a tree, resistance in an electrical circuit, wavelength in the electromagnetic spectrum or an exam percentage. A variable can be quantitative like those above or qualitative such as the colour of light.

There are two types of quantitative variable: (1) the continuous variable and (2) the discrete variable.

The **continuous variable**, for example the length of a wire being stretched, can take all values within a given range. While you may record certain discrete values, the variable (e.g. length) itself may have any value within the range of measurement.

The **discrete variable** increases in steps. The values are often whole numbers such as the number of planes in an airport which can be 1, 2, 3, . . . etc. but not 2.6 or 14.75.

Population

The population, in the statistical sense, is everything or everybody in the group you are studying. If you were selecting data about the diameters of components on a production line, the population would be the **diameters of all the components**. If you were collecting data about the leisure activities of 19-year-old students in a college, the population would be **all the 19-year-old students in that college**.

Sample

It would be impossible to collect data on a whole population. A sample is that part of the population on which observations are made. The sample must be representative of the whole population.

Statistic

A statistic is a piece of information to which we give a numerical value, for example, the arithmetic mean.

4.3 Surveys, samples and questionnaires

A survey needs careful planning. Five steps are outlined below.

1. State the nature of the task
    For example, for a survey of students decide what you mean by 'student': full-time, part-time, mature? Does a student who attends two different departments need to be counted once or twice?

2. Select the sample
    A representative sample is usually randomly chosen but it may be better to subdivide the population and take a random sample in each part. A random sample is one in which each member has the same chance of selection. It could be selected by allotting numbers to members and choosing random numbers 'out of a hat' or with a calculator or computer. For example a college roll could be used as a numbered list of students.

    The larger the number of members in the sample, the greater is the chance that the results will be valid.

3. Construct a questionnaire
    Decide what information you need to know. Phrase the questions carefully. Your aim is to obtain reliable unambiguous answers which can be tabulated. The principles are the same whether you are dealing with things or with people. Here are some essential points:

    * Each question is short and relates to one point only.
    * It should not give offence or be emotionally loaded.
    * It should not suggest the answer. [e.g. 'What do you think of the **splendid** new sports hall?']
    * The number of questions should be kept to a minimum.
    * People may prefer not to give their names.

4. Run a pilot survey
    Try out your questionnaire on a small sample to see if it furnishes the information you need.

5. Collect the data.

4.4 Classifying and tabulating data

To collate the mass of information, you need to design a table that will display it meaningfully. Enter data methodically. You need to be able to read the completed table quickly.

Table 4.1 shows the results of asking students how they travelled to college. [Your questionnaire should have been designed to cover whether walking included the distance to and from the bus, etc.]

Table 4.1 Survey of modes of travel to college

Mode of travel	Number of students
Bus	45
Train	40
Walk	35
Cycle	25
Car	20
Motor bike	10
Other	3

A **tally chart** may help with counting before tabulating. The count is made by recording a mark for each and grouping the marks in fives as shown in Table 4.2. The survey was conducted to find the kind of vehicle passing a certain point on a road over a specific time period.

Table 4.2 Record of vehicles passing in one particular hour

Type of vehicle	Tally	Total
Cars	ЖŤ ЖŤ ЖŤ ЖŤ |||	23
Buses	ЖŤ		6
Small lorries	ЖŤ ||	7
Motor bikes	||||	4
Heavy vehicles	|||	3
Other	||	2

Table 4.2, with or without the tally, is called a frequency table. The number in the final column is the **frequency** of the event recorded. The pattern which emerges is called the **frequency distribution**.

Data can be grouped as above to make them more meaningful. For example, a large number of observations are made of heights to the nearest centimetre of female students, aged 19, in the college. The individual measurements convey little when the list is read, but if the values are grouped, the pattern is clear. Table 4.3 and Fig. 4.2 show the result. The table gives the frequency distribution which is displayed on the **bar chart**.

Table 4.3 Heights of female students

Height/cm	150–154	155–159	160–164	165–169	170–174	175–179	180–184
Number of students	2	6	11	12	7	6	1

Figure 4.2 Heights of female students

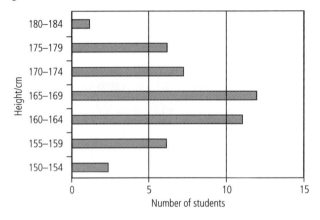

EXAMPLE

The number of students in a college cafeteria at a certain quiet time of day, for each of 28 days, is as follows:

Number of students						
3	7	11	13	12	2	16
10	21	13	2	14	6	8
12	17	9	11	16	15	1
13	4	22	14	18	5	6

Using intervals of 0–5, 6–10, 11–15, 16–20, 21–25 make a tally chart to obtain the frequency distribution.

The results are tabulated as a frequency distribution in Table 4.4.

Table 4.4 Count of number of students in the cafeteria

Number of students	Tally Number of days	Frequency
0–5	ⅢⅠ	6
6–10	ⅢⅠ	6
11–15	Ⅲ Ⅲ	10
16–20	‖‖	4
21–25	‖	2

EXERCISE 4.1 Tally charts and frequency tables

1. Thirty telephone calls timed to the nearest minute are given below:

Minutes									
18	4	7	9	2	15	4	3	12	7
1	20	5	10	11	12	14	6	20	5
6	19	2	13	2	3	9	17	15	4

Group them and arrange them in a frequency table.

2. During one lunch hour, a tally was taken of the main dishes chosen by 76 students. 21 ate beefburgers, 25 ate chicken and chips, 19 ate sandwiches and 11 ate beans on toast.

Draw the tally chart and make a frequency table for these data [look at Table 4.2].

4.5   Charts and diagrams

On its own, the frequency distribution in Table 4.3 conveys little, but the bar chart in Fig. 4.2 makes the information clear. We shall look at six ways of presenting data pictorially.

Pictograms

A pictogram uses a symbol to represent number and the symbol is used repeatedly to represent the data as a whole.

EXAMPLE

Figure 4.3 Pictogram of student intake

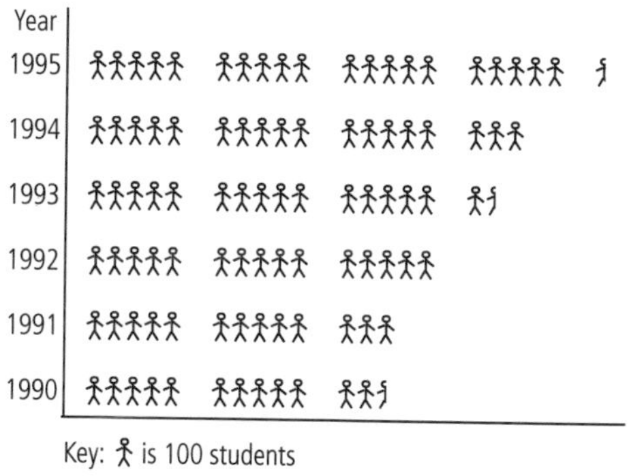

Key: ☺ is 100 students

Figure 4.3 shows how student intake varied in the early years of a new college. Each symbol represents 100 students. Data can be shown to the nearest 50. How many students were there in the year of opening?

1250

What trend does the pictogram display?

Increasing student members, year on year.

EXERCISE 4.2 Interpreting a pictogram

1. Again, use Figure 4.3.
   (a) What was the student intake in 1995?
   (b) When did the intake exceed 2000 students?
   (c) Between which years did the smallest expansion occur?

Bar charts

Data can be represented by a series of bars all of the same width, arranged horizontally or vertically. The bars may or may not have gaps between them. Each bar is labelled and the length or height gives the frequency. The scale must be stated. Bar charts can show qualitative or numerical data.

EXAMPLE

College engineering courses were surveyed to show how the students were distributed amongst the specialist subjects. The data is in Table 4.5 and displayed in Fig. 4.4. Which engineering subject is most popular?

Motor Vehicle Engineering

Table 4.5 Student members in engineering courses

Subject	College course	Frequency
Mechanical	1	20
Production	2	12
Electrical	3	15
Chemical	4	15
Motor vehicle	5	25

Figure 4.4 Bar chart of student numbers on courses 1 to 5

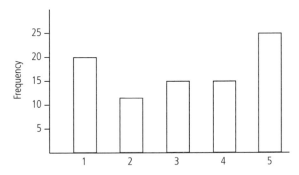

Other examples of bar charts are illustrated in Fig. 4.5.

Dual bar charts present two sets of similar data which can be compared as in Fig. 4.6. The dual bar chart shows that department A's profit has recently declined while department B's has increased steadily to outpace department A.

Sectional bar charts are used to display information on the same topic which can be divided into components. Fig. 4.7 shows the student numbers divided into full and part-time students.

Figure 4.5 More examples of bar charts

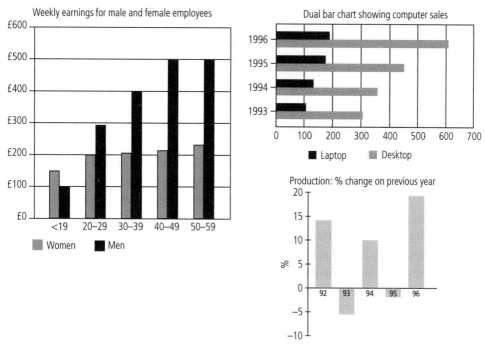

Figure 4.6 Dual bar chart: profits in departments A and B

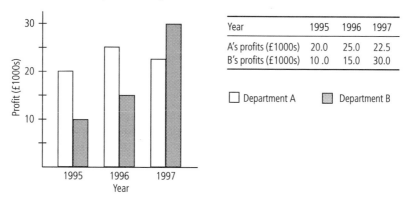

Year	1995	1996	1997
A's profits (£1000s)	20.0	25.0	22.5
B's profits (£1000s)	10 .0	15.0	30.0

☐ Department A   ▨ Department B

Figure 4.7 Sectional bar chart: part-time and full-time student numbers

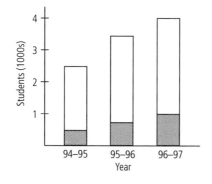

Year	1994–5	1995–6	1996–7
Part-time (1000s)	2.50	3.50	4.00
Full-time (1000s)	0.50	0.75	1.00

☐ Part-time   ▨ Full-time

EXERCISE 4.3 Bar charts

[You will need graph paper to draw the charts]

1. The number of goals scored in 20 football league matches over a certain Saturday were:

Number of goals	0	1	2	3	4
Number of matches	1	4	3	10	2

Draw a vertical bar chart to represent these data.

2. The types of transport used by 100 people to get to work on a certain day were:

Type of transport	Rail	Bus	Cycle	Car
Number of people	40	20	5	30

Draw a bar chart using **horizontal** bars to represent these data.

3. The number of part-time and the number of full-time students in Further Education colleges in the UK over a period of three years were:

College years	1994–95	1995–96	1996–97
Part-time (millions)	3.9	4.5	4.9
Full-time (millions)	0.75	0.8	0.9

Represent these data as accurately as you can by drawing a sectional bar chart.

4. Select a sample, half female half male, of students (or friends and family) who have birthdays in the first half of the year.

Draw a table as shown and then represent these data on a dual bar chart.

Birthday month	Jan	Feb	March	April	May	June
No. of females						
No. of males						

(Think carefully of the size of your sample).

Pie charts

A pie chart is a circle which represents the whole of the data. It is divided into sectors such that the area of each sector is proportional to the data represented.

To draw a pie chart using given data you need to calculate the angle θ for each sector in Fig. 4.8.

Area of sector represents data in that sector.
Area of circle represents the whole.

Thus $\dfrac{\theta}{360°} = \dfrac{\text{area of sector}}{\text{area of circle}} = \dfrac{\text{data represented}}{\text{total}}$

$$\Rightarrow \theta = \frac{\text{data represented}}{\text{total}} \times 360°$$

Figure 4.8 A pie chart

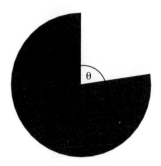

$$= \text{fraction of whole} \times 360°$$

If the chart displays percentages:

$$\theta = \frac{\text{percentage represented}}{100} \times 360°$$

EXAMPLE

90 students were asked to name their one major sports activity. Represent the results by a pie chart.

Fitness training	Swimming	Athletics	Team games	Walking	Other sports	None
20	4	12	20	6	10	18

First, calculate the angles.

For **None** angle θ = fraction of whole $\times 360°$

$$= \frac{18}{90} \times 360°$$

$$= 72°$$

The other angles have been calculated:

Fitness training	Swimming	Athletics	Team games	Walking	Other sports	None
80°	16°	48°	80°	24°	40°	72°

The pie chart is shown in Fig. 4.9.

Figure 4.9 Sports activities

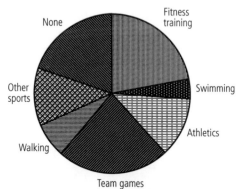

EXERCISE 4.4 Pie charts

1. The courses studied by 120 students are:

Leisure and tourism	Health and social care	Engineering	Business
40	35	18	27

Represent the data by a pie chart.

2. A business spends £36 000 on overheads, $\frac{1}{2}$ is on wages, $\frac{1}{4}$ on rent, $\frac{1}{8}$ on utilities and $\frac{1}{8}$ on other items. Draw a pie chart to represent the expenditure.

3. A television manufacturer produces 90% of its output in black, 7% in white and 3% in other colours. Show this information on a pie chart.

4. The group sales of an international company are given as percentages of the total which is £60 million. The data are given on the pie chart in Fig. 4.10.

Figure 4.10 Worldwide sales

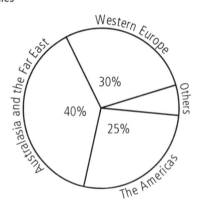

(a) Find the angle of each sector by calculation.
(b) From the angles calculate the sales figures in £ for each sector and draw up a table giving the results.

Line graphs

Line graphs show trends more clearly than bar charts. Also, it is easier to add data to a line graph. They are commonly used to present records of variables which are measured at different times. Examples include temperature, rainfall, production, blood pressure, etc.

EXAMPLE

The average daily maximum temperature in °F at the holiday resort of Chiang Mai in Thailand over a five month period is as follows:

Month	Jan	Feb	March	April	May
Temp. °F	84	89	94	97	94

Draw a line graph to represent these data.

Figure 4.11 Temperature in Chiang Mai

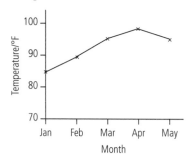

We must think carefully when choosing the scale for temperature. It would not be a good idea to start at 0°F. Can you see why not?

The above example emphasizes the importance of choosing an appropriate scale and observing carefully any scale given. A line graph of blood pressure with time, for example, could be made to look very frightening if it had high peaks and low troughs, or too reassuring if there appeared to be too little variation. As we said at the beginning of the chapter, the disguised zero can mislead.

Also, we have discussed the need to choose carefully, or to examine, the scale along the vertical axis.

In order to avoid misrepresentation, it is common to start at 0 and then make a zigzag line to draw attention to the shortening of the scale. An example is given in Fig. 4.12.

Figure 4.12 Rising expenditure (vertical scale shortened)

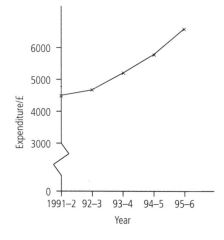

EXERCISE 4.5 Line graphs

[You will need graph paper.]

1. The number of students doing each of 5 courses is as follows:

Course number	1	2	3	4	5
Number of students	20	12	15	15	25

   (a) Draw a line graph to illustrate the figures.
   (b) Now refer back to Fig. 4.2. Which method of presentation do you find preferable in this case – a line chart, a pie chart or a bar chart? Why?

2. The average daily maximum temperature and the average daily minimum temperature in °F at Kathmandu, Nepal for the months January to May are:

Month	Jan	Feb	March	April	May
Max. temp.	65	67	77	83	86
Min. temp.	35	39	45	53	61

   Represent these data by drawing two line graphs, one for the average maximum temperature and one for the average minimum temperature on the same axes.

NOTE: The line graphs drawn to display statistical information have points joined by straight lines. They are not representing continuous variations like the graphs you will study in Chapter 7.

Histograms

A histogram looks similar to a bar chart but there are important differences. They are listed in Table 4.6.

Figure 4.13 provides some illustrations.

When we are constructing a histogram from data we must first decide how many class intervals we need and so what size they will be. Calculate the range of the data, and judge how many classes will give reasonable accuracy in displaying the results. The fewer classes there are, the greater is the chance of misrepresenting the figures. Too many intervals may have too few observations in each. Decide also whether some class intervals should be

Table 4.6 Differences between bar charts and histograms

Feature	Bar chart	Histogram
Data	Discrete values	Spread of values
Horizontal axis	Labelled for each bar	Continuous scale
Class size	Need not be equal	Intervals need not be equal
Bar width	Equal	Appropriate to the class interval
Frequency	Height of bar	Area of bar
Bar separations	Often with gaps	Usually without gaps and sometimes without verticals
Sign	Can show + and –	Always positive

Figure 4.13 Bar charts and histograms

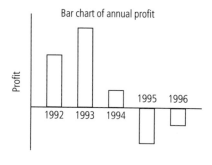

Bar chart of annual profit

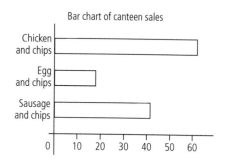

Bar chart of canteen sales

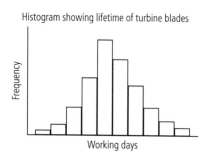

Histogram showing lifetime of turbine blades

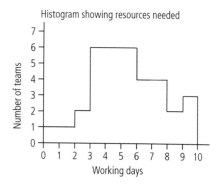

Histogram showing resources needed

wider than others. Your judgement will improve with experience. Usually five classes is too few and fifteen too many

[NOTE: Classes are sometimes called groups, blocks or bins.]

Next, the mid-points or the boundaries of the intervals must be calculated and located on the horizontal scale. Then the frequency bars can be drawn in.

Remember that the horizontal scale is continuous and each interval starts exactly where the previous one ends.

EXAMPLE

An examiner marks the papers of 50 students with results ranging from 20% to 98%. He chooses class intervals of 10% and constructs a frequency distribution table from a tally. The final marks are expressed to the nearest 1%. From Table 4.7 he constructs a histogram as shown in Fig. 4.14. The mid-points of the intervals are at 25, 35, etc.

NOTE: In more advanced work, while the marks are expressed as whole number percentages, 20 means between 19.5 and 20.4. Boundaries are then drawn at 19.5, 29.5, 39.5, etc. in the histogram.

The finished histogram shows the distribution and reveals that:

1. the most frequently scored marks are around the centre

2. very high and very low marks are rare

3. marks are more bunched together in the lower half of the distribution.

Table 4.7 Frequency distribution of exam percentages

Percentage	Tally	Frequency
20–	||	2
30–	﷼ ||||	9
40–	﷼ ﷼ ||	12
50–	﷼ ﷼ |||	13
60–	﷼ ||	7
70–	||||	4
80–	||	2
90–100	|	1

Figure 4.14 Histogram of exam results with frequency polygon

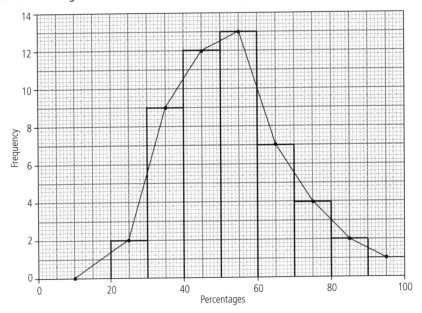

Frequency polygons

A frequency polygon is a line graph which conveys the same information as a histogram but shows more clearly the character of the distribution. Thus it makes comparison of distributions easier.

Referring again to Fig. 4.14, you will see the frequency polygon has been superimposed on the histogram. A frequency polygon can be drawn by plotting a line graph of frequency against mid-point of the class interval. It can stand on its own or it can be superimposed on the histogram as in Fig. 4.14.

EXERCISE 4.6 Histograms and frequency polygons

1. After a certain number of working hours a machined part has to be replaced owing to wear.

Records are kept of the number needing replacement as time goes on. In a sample of 100, the lifetime *L* and the number *n* discarded in each period are given in the frequency table.

L in years	0–	2–	4–	6–	8–	10–12
n	34	25	16	12	9	4

Draw a histogram to show the data.

2. The times taken to assemble a component in a factory were recorded to the nearest three seconds (i.e. 0.05 minutes). They are:

			Minutes			
2.35	2.85	2.75	4.50	2.10	3.55
3.80	3.50	3.15	2.90	2.45	3.30
4.20	3.35	3.40	3.05	3.45	3.15
2.45	2.50	2.60	3.25	3.60	3.25
3.00	2.00	2.95	3.70	2.80	3.90
2.90	3.80	3.50	3.95	2.70	3.30
1.65	4.05	3.65	2.80		

Use intervals 1.55–1.94, 1.95–2.34, etc., finishing with 4.35–4.75.

(a) Use the data to plot a histogram.
(b) Draw a frequency polygon on your histogram.
(c) Plot a separate frequency polygon.
(d) Comment on the result shown by the frequency polygon.

4.6   The normal distribution

The data in the last question are typical of a set of measurements which vary about a mean value. Can you see that if a very large number of observations are analysed with small class intervals the frequency polygon comes closer to a curve?

A graph of frequency against measurement is called a frequency curve. If the distribution is of a large number of random occurrences, the curve has a characteristic 'bell' shape and is symmetrical about the arithmetic mean (see Section 4.8).

The bell-shaped curve (Fig. 4.15) represents most natural distributions from the number of peas in a pod to the general intelligence of the population.

We call such a frequency curve a **normal distribution**. By sampling, we can find a constant for a particular set of measurements called the **standard deviation** σ which is a measure of the width of the curve. Making a larger number of measurements reduces the width of the curve and the standard deviation. Thus the value of the mean is more precisely known.

Figure 4.16 illustrates the point: as the number of observations increases, the height increases, the standard deviation decreases and the spread decreases.

Figure 4.15 The normal distribution

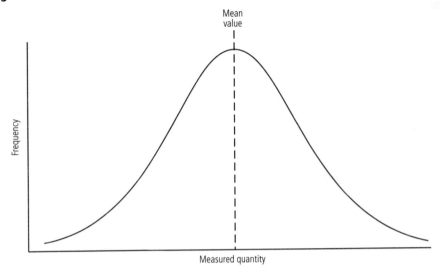

Figure 4.16 Effect of a large number of observations on the normal distribution

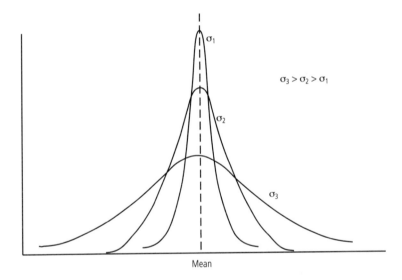

Knowing the standard deviation σ we can predict the probability of measurements being within a certain range about the mean as follows:

Range	% of observations
Within $1 \times \sigma$ of mean	68
Within $2 \times \sigma$ of mean	95
Within $3 \times \sigma$ of mean	99.7

Thus 99.7% of the data will be within the range of mean value $\pm 3\sigma$.

4.7 The cumulative frequency curve

The cumulative frequency of a distribution is the sum of the frequencies up to a certain value of the measured quantity. The cumulative frequency (i.e. the running total) increases as the value increases through its range.

We shall illustrate the point by extending the investigation of examination percentages in Fig. 4.14 and Table 4.7.

Table 4.8 includes the cumulative frequencies at the upper class boundaries. For example, for percentages 0 to 40 we have added 2 and 9 to give 11.

Table 4.8 Cumulative frequencies for exam results

Frequency table		Cumulative frequency table	
Percentage	Frequency	Percentage	Cumulative frequency
		(upper class boundary)	
20–	2	19.5	0
30–	9	29.5	2
40–	12	39.5	11
50–	13	49.5	23
60–	7	59.5	36
70–	4	69.5	43
80–	2	79.5	47
90–100	1	89.5	49
		100	50

A cumulative frequency curve is a graph of **cumulative frequency against the corresponding percentage intervals**. The points are joined by straight lines. The zero point is included for less than 20%. The points are joined by straight lines, so the graph is not strictly a curve. The 'curve' for the examination data is shown in Fig. 4.17.

NOTE: See note on p. 59. In more advanced work the upper class boundaries 19.5, 29.5, etc., would be used.

From the curve we can find

1. the median (see Section 4.8) by finding the percentage with a cumulative frequency of 50, i.e. half-way up.
2. the number or percentage of candidates who had a mark below or above a certain level.

EXERCISE 4.7 Cumulative frequency

[Graph paper will be needed]
1. Draw a cumulative frequency graph from the data in Table 4.8 using a larger scale than shown in Fig. 4.17. Find
   (a) the median mark
   (b) the number of students who failed, scoring less than 38%
   (c) the number of students who were awarded a distinction, scoring more than 75%.

Figure 4.17 Cumulative frequency curve

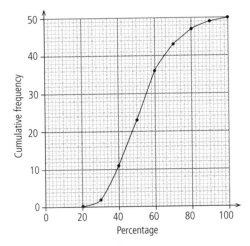

2. The cumulative frequency curve can help to identify the probability of components being defective. Resistors of 4.7 kΩ, say, could be targeted. A large number of measurements on them could give results over a range greater than ±10% of 4.7 kΩ, the stated tolerance. Results of resistance measurements of 100 components are as follows:

Resistance/kΩ	−4.0	−4.2	−4.4	−4.6	−4.8	−5.0	−5.2	−5.4
Frequency	3	8	15	24	25	14	9	2

(a) Draw the cumulative frequency curve. [You need a new table.]
(b) Calculate 4.7 kΩ + 10% and 4.7 kΩ − 10%.
(c) From the graph find the corresponding levels on the curve.
(d How many components per 100 lie outside the tolerance range?

[NOTE: Tolerance is covered in Chapter 3, Section 3.4.]

Quartiles and percentiles

Suppose we divide the distribution into four parts as shown in Fig. 4.18.

The measured value corresponding to the point $\frac{1}{4}$ of the way up the cumulative frequency range is called the **lower quartile**. 25% of the items in the distribution are below that value.

The measured value $\frac{3}{4}$ of the way up the cumulative frequency range is called the **upper quartile**. 75% of the items are below that value.

The difference between those two values is called the **interquartile range**. The middle 50% of the items in the distribution are within that range. A large interquartile range indicates a large spread in the distribution.

Similarly, we would divide the distribution into 100 parts, called **percentiles**. The 44th percentile would be the measured value where the cumulative frequency is 44% of the total number of observations, as in Fig. 4.18.

The 'curve' can be used to find how many of the batch may have to be discarded as we did in question 2 above. Go back to your graph and read off the percentiles at the tolerance limits.

Figure 4.18 Quartiles and percentiles

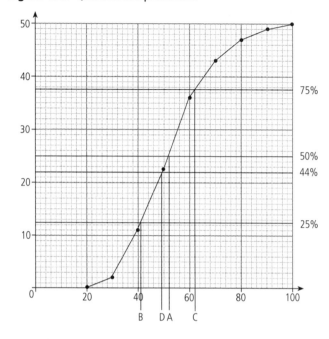

A, **median** = 52
B, **lower quartile** = 41
C, **upper quartile** = 62
Interquartile range = 62 − 41
= 21
D, **44th percentile** = 49

4.8 **Averages and range**

It is often useful to compare a **single statistic** from one set of data with a statistic from another set. The statistic used must be representative of the distribution which implies that it is near the centre of the set. Such a statistic which is a measure of the centre of a distribution is called an **average**. The most common averages in use are the **mean**, the **median**, and the **mode**. Which is more appropriate depends on the circumstances.

The arithmetic mean

The most frequently used average is the arithmetic mean (or just 'mean'). It is calculated from:

$$\text{mean} = \frac{\text{sum of the values}}{\text{number of values}}$$

Although there are situations where the mean value is not the appropriate average, it usually is the best one to use.

Consider now a situation where the mean is not representative of the distribution. The mean weekly wage of 10 workers is £190. That is a representative value as they all earn similar amounts.

Now, let us include the managing director who earns £1000 per week. We now have 11 workers.

$$\text{Mean weekly wage} = £\frac{1900 + 1000}{11}$$
$$\approx £263$$

The inclusion of the one very high wage distorts the mean and makes it unrepresentative of the wages of most workers.

Similarly, a very low value can distort the mean.

EXAMPLES

1. A student scores 40%, 30%, 65%, 70%, 49% in five tests. Find his mean percentage.

$$\text{Mean} = \frac{\text{sum of values}}{\text{number of values}}$$

$$= \frac{40 + 30 + 65 + 70 + 49}{5}\%$$

$$= 51\% \text{ (to nearest 1\%)}$$

2. The exam marks scored by 50 students were:

Percentages:

22	25	30	31	33	33	35	37	38
39	39	40	40	42	42	43	43	43
44	45	47	48	49	50	51	51	52
53	54	54	55	56	57	58	58	59
61	62	64	65	66	68	69	70	73
76	77	80	81	90				

(used for Table 4.7)

Find the average percentage.

$$\text{Mean} = \frac{\text{sum of values}}{\text{number of values}}$$

$$= \frac{295 + 302 + 315 + 320 + 336 + 250 + 256 + 259 + 265}{50}\%$$

$$= \frac{2598}{50}\%$$

$$= 52\% \text{ (to nearest 1\%)}$$

[HINT: When adding a very large number of figures, do them in batches; then add the subtotals.]

3. A cricketer's batting average for five innings was 17. In four matches he scored 26, 16, 21 and 9. What was his score in the fifth innings?

$$\text{Sum of values} = \text{number of values} \times \text{mean}$$

$$= 5 \times 17$$

$$\Rightarrow \text{total for 5 innings} = 85 \text{ runs}$$

$$\text{score in fifth innings} = 85 - (26 + 16 + 21 + 9)$$

$$= 13 \text{ runs}$$

4. Find the mean height of a female student from the frequency distribution (from Table 4.3).

Height/cm	152.5	157.5	162.5	167.5	172.5	177.5	182.5
No. of students	2	6	11	12	7	6	1

Sum of heights $= (2 \times 152.5) + (6 \times 157.5) + $ etc. $= 7502.5$

Number of values $= 2 + 6 + 11 + $ etc. $= 45$

Arithmetic mean $= \dfrac{7502.5}{45}$ $= 166.7$ cm (4 s.f.)

EXERCISE 4.8 Arithmetic mean

1. A booklet has 14 pages and contains 6300 words. What is the average number of words per page?
2. Three packs of identical small electrical components have contents of 25 parts, 26 parts and 29 parts, respectively. Find the mean number per pack.
3. Petrol prices at five different garages are 61p, 60p, 63p, 62p and 65p per litre. Find to the nearest 1p the average price per litre.
4. Find the mean from the following frequency distribution.

Resistance/kΩ	3.9	4.1	4.3	4.5	4.7	4.9	5.1	5.3
Frequency	3	8	15	24	25	14	9	2

The resistance figures are the mid-points of the class intervals.

The median

The median of a set of values is the middle one (or the arithmetic mean of the two middle ones) if the values are listed in order of size.

The median is the average usually chosen to represent a set of examination results or a set of salaries.

EXAMPLE

Find the median of the numbers 21, 14, 17, 6, 33, 15.

Listed in increasing order: 6, 14, 15, 17, 21, 33.

The median is the mean of the two middle numbers

i.e. median $= \dfrac{15 + 17}{2} = 16$

The mode

The **mode** is the most frequently occurring value in the set. The mode is useful for shopkeepers. It can identify the 'best-seller'.

In some situations there is no mode. In others, there may be more than one mode.

EXAMPLES

1. The set 2, 2, 5, 7, 9, 10, 10, 10, 11, 12 has **mode 10**.
2. The set 2, 3, 5, 8, 12, 15 has **no mode**.
3. The set 1, 3, 5, 5, 5, 6, 7, 8, 8, 8 has **two modes**: 5 and 8. It is said to be **bimodal**.

Modal class

If values are grouped, the **modal class** is the group with the highest frequency. The modal class of the frequency distribution of exam percentages from Table 4.7 is **50 to 59**.

Percentage	20 to 29	30 to 39	40 to 49	50 to 59	60 to 69	70 to 79	80 to 89	90 to 100
Frequency	2	9	12	13	7	4	2	1

EXERCISE 4.9 Median and mode

1. A shopkeeper recorded shoe sales. The following sizes sold: 7, 8, $7\frac{1}{2}$, 5, 6, 9, 8, 5, 9, $4\frac{1}{2}$, 5. What was the modal shoe size that day?

2. Unleaded petrol costs 59p, 61p, 60p, 63p, 61p, 65p, 59p and 62p at different garages.
   (a) What is the modal cost?
   (b) What is the median cost?

3. The salaries of nine employees are: £13 500, £12 000, £15 000, £11 500, £17 500, £12 000, £11 000, £12 500, £18 000. What is the median salary?

Range

We have seen that an average can be used as a statistic to represent a set of data. Alone, however, it gives no indication of the spread of values. The **range**, i.e. **the difference between the highest and lowest values** is the simplest measure of spread.

For example, the temperature in Antalya, Turkey, falls to 4°C in January and rises to 29°C in September. The range is therefore 25°C.

This example illustrates the shortcoming of the range as defined. It measures the span of the values but in many applications it is necessary to know the lowest and highest values. In engineering the range is often expressed in that way. For example 'the range of values of furnace temperature is from 1200°C to 1400°C' is more useful then 'the range is 200°C'.

4.9 Probability

In mathematics, probability has a precise meaning. Probability is a **measure of likelihood** that an event will happen. We often use probability to predict an outcome.

Scale of probability

The scale of probability is from 0 to 1.

If an event is impossible: probability = 0.
If an event is certain to occur: probability = 1

Fractions between 0 and 1 measure probabilities of uncertain events.

Probability of an event

$$\text{Probability } P = \frac{\text{number of ways of getting event}}{\text{total number of possibilities}}$$

EXAMPLES

Event	Probability
Getting a tail from toss of a coin	$\frac{1}{2}$
Drawing a heart from a pack of cards	$\frac{1}{4}$
Sun rising tomorrow	1
Death	1 (a certainty)
Winning the lottery	$\frac{1}{14\,000\,000}$
Tomorrow is Tuesday	$\frac{1}{7}$
Today is the first of the month	$\frac{12}{365}$ (normal year)

NOTE: Probability that today is *not* the first of the month

$$= \frac{353}{365} \text{ (normal year)}$$

The last two answers must add to 1 because **1 must be the total probability of all possible outcomes**.

Similarly, when tossing a coin, probability of 'heads' + probability of 'tails' = 1

Probability from theory

We can predict the probabilities above by logical thought. There are many other examples.

Probability from experiment

We can repeatedly try to see if an event will occur. The more times we try, the more likely we are to find the true probability.

$$\text{Experimental probability} = \frac{\text{number of favourable outcomes}}{\text{total number of tries}}$$

For example if we record results of tossing a coin, we may get 'heads', say 48 times out of 100. The more we repeat the experiment, the nearer the ratio

$$\frac{\text{number of heads}}{\text{number of tosses}}$$

approaches $\frac{1}{2}$.

Random

Random means that all possibilities are equally likely. For example a card chosen at random from a pack could be any one of the 52. Random motion means without any pattern. A normal distribution about the mean is a result of observing a large number of random events.

Simple probabilities

We can calculate probability using the formula for P given earlier.

EXAMPLES

1. In a class of 15 students, two are more than 6 ft tall. If two students are selected at random from the group, what is the probability that one of them is taller than 6 ft?

   The probability is $\dfrac{2}{15}$

2. A box contains ten resistors including three 5 Ω resistors and four 8 Ω resistors. What is the probability of selecting at random
   (a) a 5 Ω resistor?

   $\dfrac{3}{10}$

   (b) an 8 Ω resistor?

   $\dfrac{4}{10}$

EXERCISE 4.10 Simple probabilities

1. A die numbered 1 to 6 is thrown. What is the probability of throwing
   (a) a six?
   (b) an odd number?

2. (a) What is the probability of drawing the ace of spades from a pack of 52 cards?
   (b) What is the probability of drawing a heart from a 52-card pack?

3. In a box of screws, 5% are defective. If one is selected at random, what is the probability that it is
   (a) defective
   (b) perfect?

4. A packet of capacitors has 8 odd ones mixed in with the normal contents, of which there are 37 left. What is the probability of selecting at random
   (a) one of the right kind
   (b) one of the odd ones?
   (c) If, first, one draws an odd one and puts it aside, what now is the probability of drawing a good one?

Mutually exclusive events and the addition rule

Events are said to be **mutually exclusive** if **either** one **or** the other can happen but **not both**. The probability of one **or** the other occurring must be more than the probability of one occurring. So for two events A and B

Probability of **A or B** $= P_A + P_B$

where P_A is the the probability of A only
and P_B is the probability of B only.

EXAMPLE

What is the probability of throwing a two or three with a six-sided die?

$$P_2 = \frac{1}{6} \quad P_3 = \frac{1}{6}$$

$$P_{2\,or\,3} = \frac{1}{6} + \frac{1}{6}$$

$$= \frac{1}{3}$$

Independent events and the multiplication rule

Two events are **independent** if one event has **no effect** on whether the other can occur.

What is the probability of two independent events A and B both occurring? The chance of both is clearly less than the chance of only one.

For two independent events A and B

$$P_{A\,and\,B} = P_A \times P_B$$

NOTE: Here you will be multiplying fractions less than 1. The answer must be smaller than either of them.

EXAMPLE

What is the probability of throwing two successive threes in two throws of a six-sided die?

$$P_3 = \frac{1}{6}$$

Probability of two successive threes $P_{3\,and\,3} = \frac{1}{6} \times \frac{1}{6}$

$$= \frac{1}{36}$$

EXERCISE 4.11 Addition and multiplication rule

[REMEMBER: use your common sense]

1. The probability of a team winning a match is $\frac{2}{3}$, based on past performance. They have three matches to play. What is the probability that they win all three?

2. What is the probability of drawing an ace or a king from a 52-card pack?

3. The probability of trains leaving a station late is $\frac{1}{12}$. The probability of subsequent delay caused by signalling problems is $\frac{1}{20}$. What is the probability of a train arriving late at its destination?

4. What is the probability of throwing a total of 10 in a single throw of two dice (six-sided)?

The probability tree

When two or more events occur, a chain of probabilities results. If a diagram is drawn showing all the likely outcomes at each stage with their individual probabilities, it is easier to find the eventual probability of a particular result. A tree is constructed stemming from the trunk on the left of the page (Fig. 4.19).

Figure 4.19 A probability tree diagram

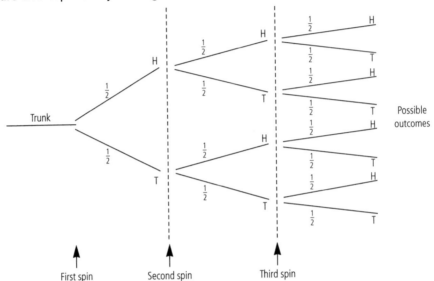

Considering again the toss of a coin, there are two possible outcomes which are drawn as branches as in Fig. 4.19 and labelled H and T. Each of the probabilities (both $\frac{1}{2}$) is marked alongside. From each, there stem two more possibilities.

After the next spin, look at the outcome. For example, what is the probability of two successive throws giving a 'tail'? You can see there is a chance of 1 in 4.

i.e. $P_{2T} = \frac{1}{2} \times \frac{1}{2} = \frac{1}{4}$

Continuing, the next spin furnishes more branches on the tree. Now, what is the probability of successively throwing a head, a tail, then a head? Again there is one path from the trunk to the devised outcome.

$P_{HTH} = \frac{1}{2} \times \frac{1}{2} \times \frac{1}{2} = \frac{1}{8}$

If, on the other hand, two heads and a tail are wanted in any order, there are three routes from the trunk. The chance of the desired outcome is 3 in 8.

$P_{2H \text{ and } T} = \frac{1}{8} + \frac{1}{8} + \frac{1}{8} = \frac{3}{8}$

The routes from trunk to outcome are through independent events so probabilities are combined using the multiplication rule. The events at the branch tips are mutually exclusive so probabilities are combined by adding.

EXAMPLE

An electronics system has two stages, A and B. The probability of failure at stage A is 0.002. The probability of failure at stage B is 0.001 (Fig. 4.20).

Figure 4.20 Probability tree

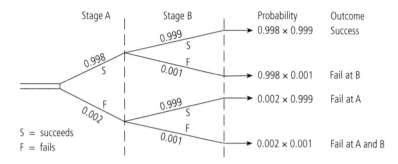

S = succeeds
F = fails

(a) What is the probabilty of the system failing?

Probability of succeeding = 0.998×0.999

= 0.997

Probability of failing = $1 - 0.997$

= 0.003

i.e. 3 in 1000 probability of failing

(b) What is the probability of both parts of the system failing?

Probability of failing at A and B = 0.002×0.001

= 0.000 002

i.e. 2 in 1 000 000 probability of both A and B failing

The same method can be applied to many problems where different probabilities are involved. In industry the method is applied to decision making. The probability of each particular outcome may be estimated experimentally. The result will provide an empirical probability which can be used on the tree.

Figure 4.21 illustrates a complex probability problem which is not considered in this book. You should be able to follow its structure. The diagram has different probabilities on different branches. Also at one stage there are three alternatives instead of two.

Figure 4.21 A more complex tree diagram

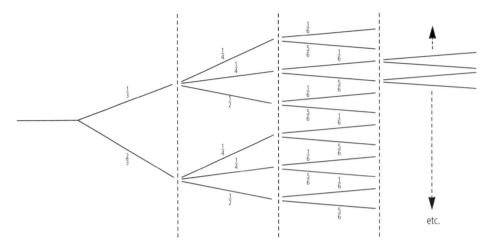

4.10 More charts which are useful in engineering

Networks

A network is a system of lines and junctions called areas or links and nodes or vertices respectively. An everyday example is the London Underground map, where the stations are nodes and the lines are links.

The aim may be to traverse the network by the shortest route or with the fewest intersections. The drawing is not to scale but the links may be labelled with distances or, perhaps, costs.

An example is illustrated in Fig. 4.22, where the shortest route from Oneville to Fiveton is needed.

Figure 4.22 A network to solve the shortest route problem

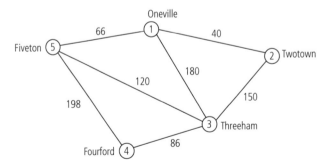

Applications include critical path analysis, minimum cost planning, maximum flow, airline scheduling, etc.

Decision trees

Decision trees are used for decision making, particularly where interdependent decisions have to be made one after another. They are constructed like networks, having nodes and branches. Like probability trees they proceed from left to right through a series of stages. Decision points where a choice of strategy has to be made are shown by square nodes. Chance nodes where the probability of different outcomes must be assessed are shown by circles. Figure 4.23 shows a simple example.

Figure 4.23 A decision tree

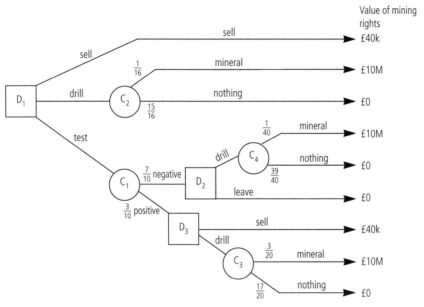

EXAMPLE

A company owns land that possibly contains mineral deposits. Do they

    (a) sell it?
    (b) drill?
    (c) test to see if mining is worthwhile?

Figure 4.23 shows the decision tree.

    Best course of action: test. If positive, drill
                              If negative, leave

The decision tree is most useful in situations where there is no exactly right answer. In everyday life, a similar problem might be 'Shall I take an umbrella?'

Flow charts

A flow chart shows a process step by step from start to finish via a number of decision boxes and loops. Actions are in rectangular boxes, decisions in diamonds. The aim may be to find the cost of a process, the time taken or the best sequence for an operation, for example.

Figure 4.24 shows a chart for finding the cost of a new car. Find the answer for your choice of car.

Figure 4.24 A flow chart: cost of a new car 'on the road'

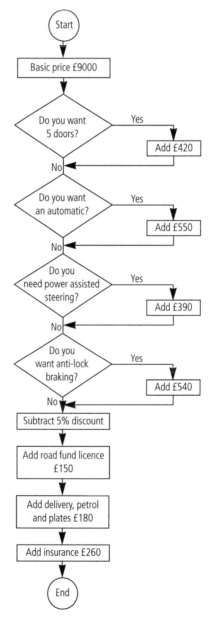

Gantt charts

A Gantt chart is a special kind of bar chart which is used in project planning. The project is broken down into tasks for which realistic estimates of time required are predicted. Some tasks are dependent on others. Certain tasks can overlap. Then a chart is constructed with tasks listed in the left hand column and with a horizontal time scale. The aim is to complete the project in the shortest possible time using the equipment and manpower resources as efficiently as possible. Table 4.9 gives an example of a planning schedule. The

Table 4.9 Planning schedule

Task	Description	Dependency	Duration (Days)	Resources (Men)
A	Prepare site	–	2	2
B	Install wiring	A	3	2
C	Build platform	B*	4	4
D	Order equipment	–	1	1
E	Assemble equipment	C and D	5	4
F	Install safety devices	E	1	1
G	Connect and test	E	2	1

*C can start when B is ⅔ completed

Figure 4.25 Gantt chart for schedule in Table 4.9

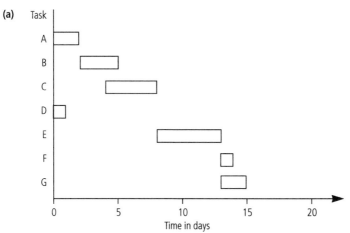

(a)

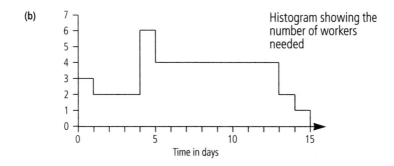

(b)

Histogram showing the number of workers needed

Gantt chart for this example is illustrated in Fig. 4.25(a). The other diagram in Fig. 4.25(b) is a histogram showing the number of workers needed in each week of the project. Overlapping the tasks may help to reduce the costs by using the workers more evenly.

To monitor progress, the bars can be coloured in. Start and finish times and overlaps may have to be modified as the work progresses.

EXERCISE 4.12 Gantt charts

1. Table 4.10 and Fig. 4.26 show how tasks in a project are scheduled and how many operations are required for each. Tasks E and G each take 2 weeks but can start at any time in the 4-week bar allotted. Similarly, D takes 3 weeks but can slot in anywhere on the 6-week bar allotted.
   (a) Draw up a table that shows the number of workers required each week through the project.
   (b) Draw a histogram to show your results.

Table 4.10 Planning schedule ['Resource' means worker!]

Task	Time	Resource
A	1	1
B	2	1
C	2	3
D	3	2
E	2	1
F	4	2
G	2	1

Figure 4.26 Gantt chart

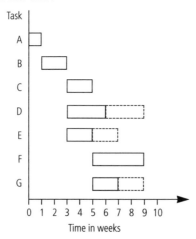

(c) Redraw the Gantt chart showing all tasks making the latest possible start. Again, use it to find the number of workers required each week.
(d) Draw a histogram for the new results. Which do you think would be a more effective use of manpower – the early start or the late start? Why?

Nomographs

Nomographs, or nomograms, are charts which show mathematical relationships. For example, if one wants to add two numbers the nomograph in Fig. 4.27 below helps.

Figure 4.27 Nomographs for adding and subtracting

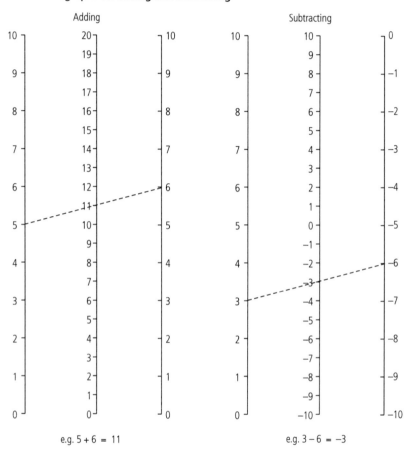

Adding

e.g. 5 + 6 = 11

Subtracting

e.g. 3 − 6 = −3

Lay your ruler across the two numbers on the outside scales. The answer appears where your ruler crosses the centre scale. Figure 4.27 also shows a similar nomograph for subtraction.

By choosing suitable scales, a nomograph can be used to multiply and divide. Charts can be made to find an answer quickly without calculation. Applications include finding the effective resistance of two resistors in parallel, calculating stress or flow rate, and also calculating many different quantities from simple mathematical relationships. Examples are shown in Fig. 4.28.

Scheduling charts

A single project involving the use of equipment, teams of workers or single operators, and comprising several tasks may need to be completed at

Figure 4.28 Nomographs

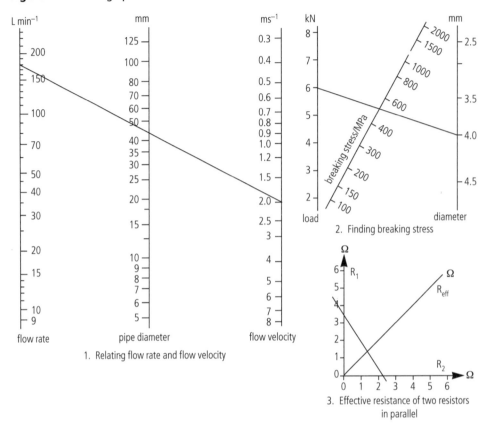

flow rate   pipe diameter   flow velocity

1. Relating flow rate and flow velocity

load

2. Finding breaking stress

3. Effective resistance of two resistors in parallel

minimum cost and in the smallest possible time span. A chart can be used to outline the plan. Figure 4.29 illustrates a typical schedule.

Figure 4.29 Workshop schedule

Machine	Day 1	Day 2	Day 3	Day 4	Day 5	Day 6	Day 7	Day 8
1								
2								
3								
4								

Job A

Job B

Job C

EXAMPLE

There are only two machine operators and Machine 3 needs to be hired by the day so should be scheduled for the shortest possible time without wasted days. Study the plan and consider the following questions.

1. For what percentage of the time is Machine 2 in use?
   25%

2. How could the schedule be arranged so that job C finishes one day earlier?
   By changing job C to day 7 and job A to day 8

3. What extra advantage would that give?
   Machine 3 would be hired for 3 days, not 4.

EXERCISE 4.13 Scheduling charts

A workshop has four machines. For a project a schedule has been planned on the chart in Fig. 4.30 and the sequence of jobs is shown in Table 4.11.

Figure 4.30 Efficient use of machines

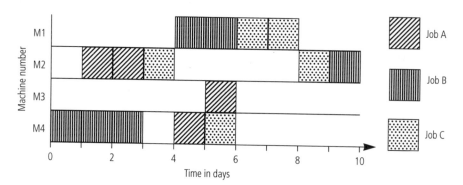

Table 4.11 Production sequence

Job A	M2 → M4 → M3
Job B	M4 → M1 → M2
Job C	M2 → M4 → M1 → M2

Consider and discuss with your group the following questions: [pencil and paper may help!]

- Could job A be completed more quickly?
- If so, what would be the effect on other jobs?
- Could job B be completed more quickly?
- What percentage of time is Machine 4 in use?
- For what percentage of the 10 days is Machine 2 non-productive?

You should have learnt that scheduling is not easy!

chapter

5

Basic techniques of algebra

5.1 What is algebra?

Arithmetic is concerned with the study of numbers. Algebra, by generalizing number using symbols, such as letters, for variables and constants helps us solve many more problems in a concise way. In algebra the letters x, y and z are commonly used as variables and other letters are used as constants. In engineering any appropriate letter may be used as either a variable or a constant.

In arithmetic the cost of buying 10 components at £3 each, for example, is found by using

$$\text{Cost} = £3 \times 10$$
$$= £30$$

Generalizing this for any number n components at £3 each the cost C in £ is found using

$$C = 3 \times n$$
$$= 3n$$

Generalizing even further there is a commonly used relationship of the form

$$y = mx$$

where y is a variable depending on the variable x and m has some constant value. In our specific example, y would be the cost, x the number of components and m the cost of each component.

Note that mx is a concise way of writing $m \times x$.

The relationship $y = mx$ is an example of an **equation**, an equation being a mathematical sentence that says that 'the left-hand side of the equals (=) sign is equal to the right-hand side'.

EXAMPLE

Consider the relationship

Force F = mass m × acceleration a in which the same mass m is acted on by different forces.

Compare $F = ma$ with $y = mx$.

$y = mx$	y (variable)	x (variable)	m (constant)
$F = ma$	F	a	m

EXERCISE 5.1 Relationships of the form $y = mx$

Copy and complete Table 5.1 as shown in the example. Think carefully as you do this.

Table 5.1 Relationships of the form $y = mx$

Equation $y = mx$ Example	y (variable)	x (variable)	m (constant)
$V = RI$	V (voltage)	I (current)	R (resistance)
(i) $s = ut$			u (constant speed)
(ii) $m = \rho v$	m (mass)	v (volume)	
(iii) $Q = mL$			L (specific latent heat)
(iv) $p = \rho gh$			ρg (density $\times$ g)

All the relationships in Table 5.1 should be familiar to you. You will note from the table that:

1. The same symbol does not always represent the same quantity. For example, in science m is used for mass but in algebra m is used as the gradient of a straight line.
2. In any relationship, different quantities can behave as variables or constants depending on the context. For example, using the pressure equation for a given liquid $p = \rho gh$, if pressure and height are the variables, then both ρ and g (the acceleration due to gravity) are constant. If, however, the same height h is chosen for different liquids, then the equation tells us how p varies with ρ.

Algebra makes use of a variety of symbols to show how expressions and variables are related to each other. Table 5.2 lists some of the basic symbols you will be using.

Table 5.2 Basic symbols used in algebra

Symbol	Meaning
$=$	is equal to
$\Rightarrow$	it follows that
$\sim$ or $\approx$	approximately
$>$	is greater than
$<$	is less than
$\neq$	is not equal to
$\ngtr$	is not greater than
$\nless$	is not less than
$\therefore$	therefore

These symbols are not new to you as you have probably used them in arithmetic problems. Use the symbols and make up examples e.g. $5 > 3$, $5 \not< 3$.

PAUSE: Think carefully about the $=$ symbol. This is the most used symbol in mathematics and yet is also the most misused! Strictly, the equality symbol links a left hand side to a right hand side and these two sides must be equal. In calculations it is sometimes neater to combine the equality sign $=$ with the 'it follows that' sign $\Rightarrow$, as in the following example.

$$P = 3 \times 60\,N \quad \text{OR} \quad P = 3 \times 60\,N$$
$$= 180\,N \quad\quad\quad \Rightarrow P = 180\,N$$

Use $\Rightarrow$ to round off a calculation but use it only where it applies.

Keeping the $=$ sign in a vertical line is a good way of working and helps to avoid errors. Avoid using equality signs all over your work in places where they are not true. There is nothing wrong with $P = 3 \times 60 = 180$ but this layout could lead to errors in more complicated algebra!

Table 5.3 gives further examples of algebraic language which you should study carefully.

Table 5.3 Algebraic language

Symbol	Meaning
$3x$	$3 \times x$ OR $x + x + x$
$2\pi r$	$2 \times \pi \times r$
ut	$u \times t$
x^2, r^3	$x \times x$, $r \times r \times r$
πr^2, $\dfrac{\pi d^2}{4}$	$\pi \times r \times r$, $\dfrac{\pi}{4} \times d \times d$
$\dfrac{1}{2}at^2$	$\dfrac{1}{2} \times a \times t \times t$

The very conciseness of the language makes accurate use essential. Far too many students confuse, for example, $2r$ with r^2 and their calculations with circles are therefore in error.

REMEMBER: r^2 is $r \times r$
$\quad\quad\quad\quad\quad\;\; 2r$ is $r + r$ or $2 \times r$

Like and unlike terms

It is important to distinguish between terms which are **like** and terms which are **unlike**. Table 5.4 gives examples of both.

Table 5.4 Like and unlike terms

Like terms	Unlike terms
a, $3a$, $-5a$	a, $2b$, $-3c$
x^2, $-3x^2$, $7x^2$	x^2, y^2, z^2
ut, $5ut$, $-3ut$	ut, ut^2, u^2t
$\dfrac{1}{2}at^2$, $3at^2$, $-5at^2$	at, a^2t, at^2

Study these carefully so that you can tell at a glance which terms are like and which are unlike in your problem solving.

5.2 Working with operators

We shall be concerned with the operators $+$, $-$, $\times$, $\div$, $\sqrt{}$, $(\ )^2$ and powers in general, on algebraic terms. The rules are the same as those for numbers, i.e. the order for $+$, $\times$ does not matter but for $-$, $\div$ it does.

Addition and subtraction of like terms

It is easy to see that like terms can be added and subtracted. For example, $3x$ (as an addition) is $x + x + x$ and $2x$ is $x + x$.

$$\Rightarrow 3x + 2x = 5x$$
$$\text{and } 3x - 2x = x$$
$$\text{while } 2x - 3x = x + x - x - x - x$$
$$= -x$$

Multiplication and division of like terms

Like terms can be multiplied and divided.

For example, $3x$ as a multiplication is $3 \times x$
$2x$ as a multiplication is $2 \times x$
$$\Rightarrow 3x \times 2x = 3 \times x \times 2 \times x$$
$$= 6x^2$$

For division $3x \div 2x = \dfrac{3 \times x}{2 \times x}$
$$= \dfrac{3}{2}$$

Addition and subtraction of unlike terms – cannot be done

You will see that unlike terms **cannot** be added or subtracted. There is nothing we can do, for example, with $3x + 2y$ or $3x - 2y$.

Multiplication and division of unlike terms

These can be multiplied and divided. For example,

$6a \times 2b$ is $12ab$; $6a \div 2b$ is $3\dfrac{a}{b}$

EXAMPLES

Collect like terms and so simplify the following expressions:
1. $2a - 3b + 5b^2 + 7a - 2b$.
   Adding and subtracting like terms the expression becomes $9a - 5b + 5b^2$

2. $-5at^2 + 3at + 2at^2 - 7a + 2at$.
   You may find it easier to arrange the like terms together in this way:

$-5at^2 + 2at^2 + 3at + 2at - 7a$

which simplifies to $-3at^2 + 5at - 7a$

[NOTE: Both answers could be further simplified by inserting brackets, discussed later.]

EXERCISE 5.2 Working with like terms

Copy and complete Table 5.5.

Table 5.5 Simplifying expressions

Expression	Simplified
1. $2u - 4u + 3$	$-2u + 3$
2. $-6t + 2t + 3t$	
3. $7t^2 + a - 2t^2$	
4. $5at + 7 - 4at$	
5. $3a - 4a + 1$	
6. $3R - 4R^2 + 2$	
7. $6t + t^2 - 6t$	
8. $\frac{3}{2}I + I - I^2$	
9. $3u \div u$	
10. $2I^2 \div I$	
11. $15t \div 3t^2$	
12. $4I \times 2I$	
13. $\frac{3a}{2} \times \frac{2}{3}a$	
14. $4I \div 2I \times I$	
15. $6x \div x^2 \times x$	

Working with powers

You know that $a \times a$ is a^2 and $a^2 \div a$ is a. Working with powers in algebra is the same as working with powers of numbers. The same rules apply.

For numbers raised to powers:

add the indices for **multiplication**
subtract the indices for **division**

In algebra the rules are the same

i.e. $a^n \times a^m = a^{n+m}$ AND $a^n \div a^m = a^{n-m}$
$\therefore a^4 \times a^2 = a^6$ $a^4 \div a^2 = a^2$

$a^{-4} \times a^2 = a^{-2}$ OR $\dfrac{1}{a^2}$ $a^{-4} \div a^2 = a^{-6}$ OR $\dfrac{1}{a^6}$

$\sqrt{a} \times \sqrt{a} = a$ OR $a^{1/2} \times a^{1/2} = a^1$

a^1 is a, a^0 is 1 (e.g. $a^2 \div a^2$ is 1)

EXAMPLE

Simplify the following:

1. $y^3 \times \sqrt{y} \div y^2$ 2. $\dfrac{(3a)^2 \times (2a)^3}{4a^4}$

1. $y^3 \times \sqrt{y} \div y^2 = y^{3^{1/2}} \div y^2$
$$= y^{1^{1/2}}$$
$$= y \times y^{1/2}$$
$$= y\sqrt{y}$$

2. $\dfrac{(3a)^2 \times (2a)^3}{4a^4} = \dfrac{9a^2 \times 8a^3}{4a^4}$
$$= \dfrac{18a^5}{a^4}$$
$$= 18a$$

EXERCISE 5.3 Multiplying and dividing powers

Copy and complete Table 5.6.

Use the working column when required.

Table 5.6 Multiplying and dividing powers

Expression	Working (if required)	Simplified
1. $u^{-1/2} \times u^{3/2} \div 3u$	$u^1 \div 3u$	$\dfrac{1}{3}$
2. $u^7 \div u^4$		
3. $a^{-3} \times 2a^3$		
4. $(2a)^2 \times (2a)^3$		
5. $u^{-2} \div u^{-2}$		
6. $(3x)^2 \div \sqrt{x} \times x^{3/2}$		
7. $(a^2)^{1/2} \times (b^3)^{1/3}$		
8. $\sqrt{x} \times \sqrt{x} \times x\sqrt{x}$		

5.3 Using brackets

BODMAS applies as it does for numbers.

Removing brackets

In algebra we often need to multiply out expressions that have brackets. In this way we **remove the brackets**. The process is also called **expanding the brackets**.

To remove or expand brackets, we need to multiply each term inside by whatever is outside.

EXAMPLES

Remove the brackets from the following:

1. $a(b+c)$ 2. $C = 20(10a+b)$ 3. $S = \frac{t}{2}(u+v)$ 4. $x-y(a-b)$

1. $a(b+c) = ab+ac$

2. $C = 20(10a+b)$
   $= 200a+20b$

3. $S = \frac{t}{2}(u+v)$
   $= \frac{ut}{2} + \frac{vt}{2}$

4. $x-y(a-b) = x-ya+yb$

REMEMBER: When you remove a bracket, multiply each term inside by whatever is outside.

Inserting brackets

Inserting brackets, called factorization, is the reverse of removing brackets. In arithmetic we factorize the number 12 by writing it as $12 = 2 \times 2 \times 3$, and 2, 2 and 3 are called the factors of 12. In algebra we factorize $ab+a^2$ by writing it as $ab+a^2 = a(b+a)$, and a and $(b+a)$ are called the factors of $ab+a^2$.

In algebra, factorization involves looking for common factors in an expression and inserting brackets. It provides another method of simplifying expressions and hence makes problem solving easier.

EXAMPLES

1. In a small electrical business certain components are bought weekly: 50 costing £a each, 5 costing £b each. What is the total cost £C for four weeks?

   There are two methods.
   $C = 4 \times 50a + 4 \times 5b$
   $= 200a + 20b$

   OR

   $C = 4 \times 5\,(10a+1b)$ [4 and 5 are common factors]
   $= 20(10a+b)$
   The second method gives a better equation (or formula) to use, given the values of a and b.

2. Write $ab+ac$ with a bracket.
   $ab+ac = a(b+c)$ [a is a common factor]

3. Simplify the formula $A = 2\pi r^2 + 2\pi rh$ by factorizing the right-hand side.
   $A = 2\pi r(r+h)$ [factors are 2, π, r, $(r+h)$]

4. Simplify the formula $V = \frac{4}{3}\pi r^3 + \pi r^2 h$

   $V = \pi r^2(\frac{4}{3}r+h)$ [Factors are π, r, r, $(\frac{4}{3}r+h)$]

EXERCISE 5.4 Using brackets

Find common factors and use brackets:

1. $xy + xa$

2. $IR_1 + IR_2$

3. $x^2y + x$

4. $3a^2 + 6ab$

5. $R = R_0 + R_0\,\alpha\theta$

Remove the brackets and collect any like terms:

6. $2a(a + b)$

7. $a^2 - a(3b + a)$

8. $x(x + y) - x(x - y)$

9. $2a - (9a + b)$

10. $x = t(u + \dfrac{1}{2}at)$

EXERCISE 5.5 More practice with brackets

Remove the brackets from the following and simplify where possible.

1. $5(a + b - 3)$

2. $7a(2 + a) - 2a(5 + 3a)$

3. $3x(x - 2) - 2x(x - 2)$

4. $-5a(2 - a + a^2)$

5. $3x(x - y) - x(2x + 5y)$

6. $-3(7 + 3I_2) + 35I_2$

7. $5(2R_1 + R_2) - 5(R_1 - 3R_2)$

8. $2t(3 + 5t) - t(-3t + 2)$

5.4 Algebraic fractions and operators

Adding and subtracting

The rules for adding and subtracting algebraic fractions are the same as for number fractions but the symbols can appear to make the calculation look more difficult than it really is.

Look at the following examples carefully.

EXAMPLES

Simplify (i.e. express as one fraction) the following fractions.

1. $\dfrac{x}{4} + \dfrac{x}{16}$ 2. $\dfrac{a}{5} + \dfrac{b}{3}$ 3. $\dfrac{3}{x} - \dfrac{2}{y}$

1. $\frac{x}{4}\left(x\frac{4}{4}\right)+\frac{x}{16}$ compare with $\frac{3}{4}+\frac{3}{16}$ (say)

$$=\frac{4x}{16}+\frac{x}{16}\qquad\qquad=\frac{12}{16}+\frac{3}{16}$$

(all fractions changed to $\frac{1}{16}$ths)

$$=\frac{5x}{16}\qquad\qquad\qquad=\frac{15}{16}$$

NOTE: You will see that, comparing both sets of fractions, x is 3, therefore $5x$ is 15, giving the same result.

2. $\frac{a}{5}\left(x\frac{3}{3}\right)+\frac{b}{3}\left(x\frac{5}{5}\right)$

$$=\frac{3a}{15}+\frac{5b}{15}$$

$$=\frac{3a+5b}{15}$$

3. $\frac{3}{x}-\frac{2}{y}$

Change these to $\frac{1}{xy}$ fractions as follows:

$\frac{3}{x}\left(x\frac{y}{y}\right)-\frac{2}{y}\left(x\frac{x}{x}\right)$

$$=\frac{3y}{xy}-\frac{2x}{xy}$$

$$=\frac{3y-2x}{xy}$$

EXERCISE 5.6 Addition and subtraction of algebraic common fractions

Simplify the following fractions:

1. $\frac{x}{3}+\frac{x}{4}$

2. $\frac{2a}{5}+\frac{b}{10}$

3. $\frac{3a}{8}-\frac{b}{16}$

4. $\frac{2}{3}y+\frac{1}{3}y$

5. $\frac{9}{x}-\frac{4}{x}$

6. $\frac{5}{x}-\frac{3}{y}$

7. $\frac{3p}{4}+\frac{5p}{16}$

8. $\frac{1}{2a}+\frac{1}{3a}$

9. $\frac{x}{3}+\frac{2x}{7}$

10. $\frac{5}{2x}-\frac{2}{3y}$

Multiplying and dividing

Algebraic fractions, like number common fractions, appear to be easily multiplied and divided since what appears to be the natural and routine way used with numbers is still correct.

e.g. $\frac{1}{x}\times\frac{2}{y}=\frac{2}{xy}$

and $\dfrac{1}{x} \div \dfrac{2}{y} = \dfrac{1}{x} \times \dfrac{y}{2}$ (invert and multiply)

$$= \dfrac{y}{2x}$$

Here are some more examples and a short exercise through which you can work quickly.

EXAMPLES

Simplify the common fractions:

1. $\dfrac{2}{5} \times \dfrac{3}{2}$ 2. $\dfrac{2x^2}{5} \times \dfrac{3}{x}$ 3. $\dfrac{a}{3} \times \dfrac{9b}{a^2}$

4. $\dfrac{a}{3} \div \dfrac{3b}{a}$ 5. $\dfrac{y}{4} \div \dfrac{y}{7}$

1. $\dfrac{2}{5} \times \dfrac{3}{2} = \dfrac{3}{5}$

(Divide top and bottom by 2)

2. $\dfrac{2x^2}{5} \times \dfrac{3}{x} = \dfrac{6x^2}{5x}$

$$= \dfrac{6x}{5}$$

3. $\dfrac{a}{3} \div \dfrac{9b}{a^2} = \dfrac{a}{3} \times \dfrac{a^2}{9b}$

$$= \dfrac{a^3}{27b}$$

4. $\dfrac{a}{3} \div \dfrac{3b}{a} = \dfrac{a}{3} \times \dfrac{a}{3b}$

$$= \dfrac{a^2}{9b}$$

5. $\dfrac{y}{4} \div \dfrac{y}{7} = \dfrac{y}{4} \times \dfrac{7}{y}$

$$= \dfrac{7}{4} \qquad \text{(dividing top and bottom by } y\text{)}$$

EXERCISE 5.7 Multiplying and dividing algebraic common fractions

Simplify the common fractions:

1. $\dfrac{3}{5} \times \dfrac{2}{3}$ 2. $\dfrac{3a}{5} \times \dfrac{2a}{6}$

3. $\dfrac{a}{7} \times \dfrac{14b}{a^2}$ 4. $\dfrac{y}{7} \times \dfrac{49b}{y^2}$

5. $\dfrac{x^3}{3} \div \dfrac{x}{3}$ 6. $\dfrac{2x}{7} \div \dfrac{3x}{14}$

7. $\dfrac{2y^2}{a} \times \dfrac{8}{y}$ 8. $\dfrac{(3a)^2}{4} \div \dfrac{3a}{1}$

9. $\dfrac{(3a)^3}{4} \div \dfrac{3a}{4}$ 10. $(-2a)^2 \div \dfrac{2a}{1}$

5.5 Substitution

A technique which you should practise is calculating the value of a collection of terms given specific numerical values which you can substitute.

EXAMPLE

Given $a = 5$, $t = 10$, $u = 3$, $s = 80$ find the value of

1. $u + at$ 2. $u^2 + 2as$

Substituting the given values

1. $u + at = 3 + 5 \times 10$ 2. $u^2 + 2as = 9 + 2 \times 5 \times 80$
$= 3 + 50$ $= 9 + 800$
$u + at = 53$ $u^2 + 2as = 809$

REMEMBER: Multiplication before addition.
$$u^2 = u \times u$$

EXERCISE 5.8 Substitution

Copy and complete Table 5.7, given that $a = 10$, $t = 3$, $r = 4$, $h = 5$. In questions 7 to 10 keep π in the answers. Set out the working for each example in the column provided if you wish.

Table 5.7 Substitution

Expression	Working	Value
1. $2a + t$	$2 \times 10 + 3$	23
2. $-5a + 3t$		
3. $3at^2 - 2t$		
4. $\dfrac{a^2}{t} - \dfrac{1}{t}$		
5. $5t - 3a$		
6. $5t + \dfrac{1}{2}at^2$		
7. $2\pi rh$		
8. πr^2		
9 $2\pi r(r + h)$		
10. $\pi r^2 h$		

5.6 More about equations

Forming equations

Take for example the equation $y = mx + c$ where c is a constant. (Note that $y = mx + c$ becomes $y = mx$ when $c = 0$.) There are many applications of this equation both in everyday life and in engineering. Table 5.8 gives a few examples. Study these carefully.

Table 5.8 Examples of $y = mx + c$ equations

Examples of $y = mx + c$	Variable y	Variable x	Constant m	Constant c
Cost of electricity/gas $C = nu + c$	Cost C	Number of units n	Cost per unit u	Standing charge c
Motion in a straight line $v = u + at$	Velocity v	Time t	Constant acceleration a	Initial velocity u
Linear expansion $l = l_0 + l_0\alpha\theta$	Length l	Temperature change θ	$l_0 \times$ coefft of expansion $l_0\alpha$	Initial length l_0

You need to have practice in forming equations. The following exercise is a mixed collection. Some equations will be of the form $y = mx$ and others will have the form $y = mx + c$.

EXERCISE 5.9 Forming equations

Form an equation for each of the following:

1. In a factory making electrical components the total cost of manufacture C is calculated by adding the overheads h to the cost of making x components at £n each.

2. The net take-home pay N is calculated by subtracting deductions D from gross pay G.

3. A father aged y years is twice as old as his son aged x years.

4. A father aged y years is 3 years more than twice as old as his daughter aged x years.

5. The perimeter P of a rectangle is twice the sum of the length l and breadth b.

6. The area of a square A is the square of its side l.

7. The work done, W, by a force F in moving an object a distance d in the direction of the force is the product of the force and the distance.

8. The voltage V across an electrical conductor is the product of the resistance R of the conductor and the current I flowing through it.

Now that you have formed equations and found out how to substitute values in them, you are ready to move on to solving them.

Solving equations

Remember always that an equation is an exact balance between what is on one side of the equality symbol and what is on the other side. Solving an equation means finding the particular value of an unknown which balances the equation.

Remember the golden rule in solving:

Whatever operation is performed on one side must be performed on the other.

We shall first consider simple algebraic equations.

EXAMPLES

1.

$$2x+7 \quad = \quad 31$$

Find the value of x in $2x+7 = 31$

Take 7 from both sides

$$2x = 24$$

Divide both sides by 2

$$x = 12$$

Equation remains balanced

CHECK: l.h.s. is $2 \times 12 + 7 = 31$
$$\Rightarrow \text{l.h.s.} = \text{r.h.s.}$$

2. In the equation $5v - u = at$, find v if $u = 10$, $a = 15$ and $t = 2$.

[keep the $=$ sign in a vertical line]
$$5v - u = at$$
$$5v - 10 = 15 \times 2$$
Add 10 to both sides
$$5v = 30 + 10$$
$$5v = 40$$
Divide both sides by 5
$$v = 8$$

CHECK: l.h.s. is $40 - 10 = 30$ r.h.s. $15 \times 2 = 30$
$$\Rightarrow \text{l.h.s.} = \text{r.h.s.}$$

With sufficient practice much of the working would be carried out in your head. A typical solution may then look like this

$$5v - 10 = 30$$
$$5v = 40$$
$$v = 8$$

3. $\quad 5x - 6 = 2 - 3x \quad$ (add $3x$ to both sides)
$$5x + 3x - 6 = 2 \qquad \text{(add 6 to both sides)}$$
$$8x = 8 \qquad \text{(divide both sides by 8)}$$
$$x = 1$$

Check this in your head.

EXERCISE 5.10 Solving equations

Solve the following equations setting out your solutions clearly. Check your solutions.

1. $3x + 2 = 11$

2. $6y = 7y + 2$

3. $2v + 3v = 14 - 2v$

4. $9x = 117$

5. $\dfrac{x}{3} + 1 = 6 - \dfrac{2x}{3}$

6. $\dfrac{3x}{5} = 15$

Find t in questions 7–10 if $a = 3$, $u = 5$, $v = 15$.

7. $at = v - u$

8. $1 + t = \dfrac{v - u}{a} + 1$

9. $u + at = v$

10. $5at = v^2 - u^2$

MORE EXAMPLES

Solve the following equations.

1. $2(x+1) + 3(x-6) = -1$
$2x + 2 + 3x - 18 = -1$ (removing brackets)
$5x - 16 = -1$ (collecting like items)
$5x = -1 + 16$ (add 16 to both sides)
$5x = 15$
$x = 3$

2. $5(x-3) - 2(x-9) = 0$
$5x - 15 - 2x + 18 = 0$ (removing brackets. Note: -2×-9 is $+18$)
$3x + 3 = 0$ (subtract 3 from both sides)
$x = -1$ (divide both sides by 3)

3. $\dfrac{1}{x} = \dfrac{5}{3}$

$1 = \dfrac{5x}{3}$ (multiply both sides by x to get x on top)

$3 = 5x$ (multiply both sides by 3)

$x = \dfrac{3}{5}$ (divide both sides by 5 and rearrange)

OR

$\dfrac{1}{x} = \dfrac{5}{3}$

$\dfrac{x}{1} = \dfrac{3}{5}$ (turn both sides upside down to get x on top!)

$x = \dfrac{3}{5}$

The second method is clearly quicker but only in an equation of this type. Look carefully at the next example.

4. $\dfrac{2}{a} + \dfrac{3}{2a} = \dfrac{1}{4}$

You must first add the fractions on the left-hand side

$\dfrac{2}{a}\left(\times \dfrac{2}{2}\right) + \dfrac{3}{2a} = \dfrac{1}{4}$

$\dfrac{4}{2a} + \dfrac{3}{2a} = \dfrac{1}{4}$

$\dfrac{7}{2a} = \dfrac{1}{4}$

$\dfrac{2a}{7} = \dfrac{4}{1}$ (turning both sides upside down)

$2a = 28$ (multiplying both sides by 7)

$a = 14$ (dividing both sides by 2)

CHECK: l.h.s. $\dfrac{2}{14} + \dfrac{3}{28}$ is $\dfrac{4}{28} + \dfrac{3}{28}$ i.e. $\dfrac{7}{28}$ or $\dfrac{1}{4}$
$\Rightarrow$ l.h.s. = r.h.s.

EXERCISE 5.11 Solving equations with brackets

1. $3(x-2) = 3$

2. $3(x+6)-14 = 7$

3. $9(2x+1)+2(x+7) = 33$

4. $2(x-3)-(x+4) = 8$

5. $4+3a = 2(a+5)+1$

6. $5-3a = 2(a-5)$

7. $\dfrac{1}{R} = \dfrac{3}{4}$

8. $\dfrac{2}{R}+\dfrac{3}{2R} = \dfrac{14}{3}$

9. $4(3a+1) = 7(a+4)-2(a+5)$

10. $5(2-3y)-4(y+1) = 44$

Yet more equations

EXAMPLE

Solve $x^2 = 36$

$\sqrt{x^2} = \sqrt{36}$ (take square root of both sides)

$x = +6 \text{ or} -6$

Check with $x = +6$ Check with $x = -6$

l.h.s. is 36   l.h.s. is (-6×-6) i.e. 36

When finding the square root of a number there will be two answers, one negative and one positive. However, in a practical context, such as engineering, you would discard the negative solution if it did not make sense.

EXERCISE 5.12 Solving and checking equations

Solve the following equations and check where appropriate. You do not need a calculator.

1. $x^2 = 121$

2. $3a^2 = 48$

3. $\dfrac{6}{a^2} = \dfrac{3}{2}$

4. $\dfrac{15}{4p^2} = \dfrac{3}{5}$

5. $\sqrt{x+2} = 3$
[HINT: Square both sides]

6. $2 \times a^{1/3} = 4$
[HINT: Cube both sides]

7. $(x+2)^3 = 27$
[HINT: Take cube root of both sides]

8. $\dfrac{x+3}{4} = \dfrac{x-3}{5}+1$

5.7 Transposing equations and formulae

A formula is in fact an equation with the relationship between quantities arranged in such a way as to enable an unknown quantity to be calculated. For example, $l = l_0(1+\alpha\theta)$, $T = 2\pi\sqrt{\dfrac{l}{g}}$ are formulae from which we could calculate l and T given all the other values. In these examples l and T are each called the **subject** of the formula. Suppose, however, that α is the unknown, as would be the case in an experiment to find α. We should then require α as the subject of the formula and in order to do this we should need to rearrange the formula. This 'changing places' of the symbols is called **transposing** the formula. You will need to do this frequently and you will find it quite straightforward if you realize that the formula is an equation.

EXAMPLE

The force F acting on a mass m and giving it an acceleration of a is given by the formula (equation) $F = ma$. Make a the subject.

$$F = ma$$

divide both sides by m

$$\frac{F}{m} = a$$

$$a = \frac{F}{m} \quad \text{(the subject should be on the l.h.s.)}$$

EXERCISE 5.13 Changing the subject of a formula

Copy and complete Table 5.9 rearranging (transposing) each formula so that the variable in brackets is the subject.

Table 5.9 Transposing formulae

Formula	Working (if required)	Solution
$p = \rho gh$ (h) (Example)	$\div$ b.s. by ρg $\frac{p}{\rho g} = h$	$h = \frac{p}{\rho g}$

1. $V = RI$ (R)
2. $C = \pi d$ (π)
3. $W = FS$ (S)
4. $A = lb$ (l)
5. $V = kT$ (k)
6. $A = \pi dh$ (h)
7. $P = I^2 R$ (R)
8. $A = \pi r^2$ (r)
9. $F = \sigma A$ (σ)
10. $\omega T = 2\pi$ (ω)

It is important that you develop speed and accuracy in solving equations and in transposing formulae.

More examples in transposing

EXAMPLES

1. Make α the subject of the formula $R = R_0(1 + \alpha\theta)$

$$R = R_0(1 + \alpha\theta) \quad \text{(remove brackets)}$$
$$R = R_0 + R_0\alpha\theta \quad \text{(subtract } R_0 \text{ from both sides)}$$
$$R - R_0 = R_0\alpha\theta \quad \text{(divide both sides by } R_0\theta\text{)}$$
$$\Rightarrow \alpha = \frac{R - R_0}{R_0\theta} \quad \text{(rearranging)}$$

2. Make 'a' the subject in $R = \dfrac{\rho l}{a}$

   a is 'on the bottom'. The secret (as in Exercise 5.11) is to arrange the 'unknown' to be 'on the top'.

   $R = \dfrac{\rho l}{a}$ (multiply both sides by a)

   $Ra = \rho l$ (divide both sides by R)

   $\Rightarrow a = \dfrac{\rho l}{R}$

EXERCISE 5.14 More examples in transposing

In the following arrange the term in [ ] to be the subject.

1. $R = R_0(1 + \alpha\theta)$ $[\theta]$ 2. $R = R_0(1 + \alpha\theta)$ $[R_0]$

3. $R = \dfrac{V}{I}$ $[I]$ 4. $T = \dfrac{2\pi}{\omega}$ $[\omega]$

5. $Q = \dfrac{C}{V}$ $[V]$ 6. $T^2 = 4\pi^2\dfrac{l}{g}$ $[g]$

7. $l = l_0(1 + \alpha\theta)$ $[\theta]$ 8. $l = l_0(1 + \alpha\theta)$ $[l_0]$

More substitution

You may need some more practice in substituting in formulae where there are brackets and powers. You will recognize some of these formulae.

EXAMPLES

1. $s = ut + \dfrac{1}{2}at^2$. Find s when $u = 40$, $t = 100$, $a = 3.5$.

   $s = ut + \dfrac{1}{2}at^2$ REMEMBER: $\dfrac{1}{2}at^2 = \dfrac{1}{2} \times a \times t \times t$

   $s = 40 \times 100 + \dfrac{1}{2} \times 3.5 \times 100 \times 100$

   $= 4 \times 10^3 + 1.75 \times 10^4$

   $= 10^3(4 + 1.75 \times 10)$ [10^3 is a common factor]

   $= 21.5 \times 10^3$

   $\Rightarrow s = 2.15 \times 10^4$ (in standard form)

2. This example will show you the use of factorizing and inserting brackets before substituting.

   $A = 2\pi\dfrac{d^2}{4} + \pi dh$. Find A when $d = 20$, $h = 100$, in terms of π.

   $A = \pi d(\dfrac{d}{2} + h)$ [πd is a common factor]

   $= \pi \times 20(10 + 100)$

   $= \pi \times 20 \times 110$

   $\Rightarrow A = 2.2 \times 10^3\pi$ (in standard form)

EXERCISE 5.15 Substituting in formulae

In each of the following formulae find the value of the subject.

1. $v = u + at$ $u = 20, a = 15, t = 30$

2. $E = I^2Rt$ $I = 3, R = 5, t = 20$

3. $V = \frac{4}{3}\pi r^3$ $r = 3$

Give the answer in terms of π.

4. $V = 2\pi r^2 + 2\pi rh$ $r = 5, h = 20$
(Factorize the formula first. Give the answer in terms of π.)

5. $k = \frac{1}{2}mv^2$ $m = 30, v = 60$

Give the answer in standard form.

Transposing with powers and roots

Formulae often contain powers and roots. You will need to be able to transpose these formulae too.

EXAMPLES

1. $V = \frac{4}{3}\pi r^3$ Make r the subject.

    $3V = 4\pi r^3$ (multiply both sides by 3)

    $\frac{3V}{4\pi} = r^3$ (divide both sides by 4π)

    $r = \sqrt[3]{\frac{3V}{4\pi}}$ (take cube root and rearrange)

2. $s = \frac{1}{2}at^2$ Make t the subject.

    $2s = at^2$ (multiply both sides by 2)

    $\frac{2s}{a} = t^2$ (divide both sides by a)

    $t = \sqrt{\frac{2s}{a}}$ (square root both sides and rearrange)

NOTE: With plenty of practice you will be able to rearrange formulae quickly and easily and omit some steps.

EXERCISE 5.16 Rearranging the subject of a formula

In each of the following formulae arrange the symbol in brackets [ ] to be the subject.

    1. $E = RI^2t$ $[I]$

2. $2as = v^2 - u^2$ [u]

3. $V = \frac{\pi}{6}d^3$ [d]

4. $T = 2\pi \sqrt{\dfrac{l}{g}}$ [l]   [HINT: Square both sides]

5. $T = 2\pi \sqrt{\dfrac{l}{g}}$ [g]

6. $k = \frac{1}{2}mv^2$ [v]

7. $d = \dfrac{Wl^3}{3yAk^2}$ [y]

8. $d = \dfrac{Wl^3}{3yAk^2}$ [l]

9. $d = \dfrac{Wl^3}{3yAk^2}$ [k]

10. $T = 2\pi \sqrt{\dfrac{M+m}{k}}$ [g]

11. $Z = \sqrt{R^2 + X^2}$ (X)

12. $Z = \sqrt{R^2 + X^2}$ (R)

It is hoped that you now understand the power of algebra in generalization and hence its usefulness in application to problems. You have seen for example how so many relationships involving physical quantities can be represented by either the general equation $y = mx$ or the even more general equation $y = mx + c$. In Chapter 7 you will see how these equations are represented graphically.

Chapter review

- Algebra generalizes number by using symbols (letters) as variables and constants.

- $y = mx$ is a general equation which can represent many relationships involving physical quantities, e.g. $F = ma$.

- $y = mx + c$ is an even more general relationship.

- Rules for generalized number are the same as for arithmetic number, i.e. rules for $+ - \times \div$, BODMAS and powers. Take special care when adding and subtracting algebraic fractions.

- Techniques include:
  - collecting like terms, e.g. $2a + 3a = 5a$
  - $\times$ and $\div$ of numbers raised to powers, i.e. $n^a \times n^b = n^{a+b}$, $n^a \div n^b = n^{a-b}$
  - inserting brackets by finding common factors, e.g. $3ab + 3a = 3a(b + 1)$

– removing brackets, e.g. $3a(b + 1) = 3ab + 3a$
– substituting values in an expression.

• An equation is a statement that the l.h.s. of an equality symbol is equal to the r.h.s., e.g. $3x + 1 = 5 - x$.
  – an equation is treated like a balance
  – solving an equation means finding the unknown, e.g. x.

• A formula is an equation rearranged so that the unknown becomes the 'subject' on the l.h.s., e.g. $F = ma$, where F is the subject, can be rearranged to give $a = \dfrac{F}{m}$ where a is the subject.

chapter

6

Space and shape

6.1　Perimeter and area

Perimeter

Perimeter is a measurement of length and so has the units of length, e.g. mm, m. The perimeter of a circle has a special name – it is called the **circumference**. Figure 6.1 shows the formulae for perimeters of some common shapes.

NOTE: When using π we usually choose the approximate value $\frac{22}{7}$, 3.142, or 3 to the nearest whole number.

Figure 6.1 Perimeters of common shapes

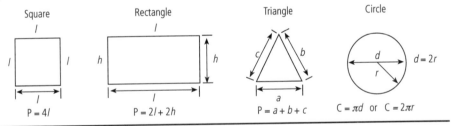

Square	Rectangle	Triangle	Circle
$P = 4l$	$P = 2l + 2h$	$P = a + b + c$	$C = \pi d$ or $C = 2\pi r$

EXAMPLES

1. Find the perimeter P of the shape shown in Fig. 6.2.

Figure 6.2 An irregular shape

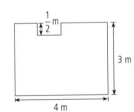

$P = (2 \times 4) + (2 \times 3) + (2 \times \frac{1}{2})\, \text{m}$

$\quad = (8 + 6 + 1)\, \text{m}$

$P = 15\, \text{m}$

2. Find the circumference C of a circle of diameter 28 mm.
   Take π as $\dfrac{22}{7}$.

   $C = \pi d$ OR $C = 2\pi r$

   $\quad = \dfrac{22}{7} \times 28\,\text{mm}$ $= \dfrac{22}{7} \times 14\,\text{mm}$

   $C = 88\,\text{mm}$ $C = 88\,\text{mm}$

EXERCISE 6.1 Calculation of perimeters

Find the perimeter of each of the shapes in Fig. 6.3. Take $\pi = \dfrac{22}{7}$.

Figure 6.3 Shapes for Exercise 6.1

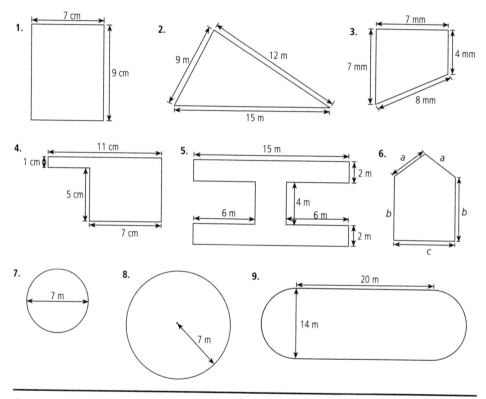

Area

Area is a measure of the size of a surface. Different shapes can have the same area, while shapes with the same perimeter can have different areas.

Calculation of area

Area is calculated by multiplying a length by a length. Units of area are length², e.g. mm², m².

It is said to have two 'dimensions'. Figure 6.4 gives the formulae for the areas of some common shapes.

NOTE: The area of a triangle $= \frac{1}{2}$ area of rectangle with sides b and h. Can you see this from Fig. 6.4?

REMEMBER: $A = \pi r^2$ means $\pi \times r \times r$

$A = \dfrac{\pi d^2}{4}$ means $\dfrac{\pi}{4} \times d \times d$

Figure 6.4 Formulae for area

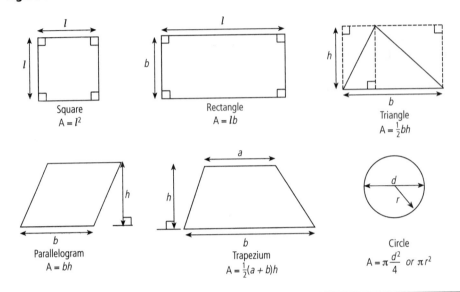

Square
$A = l^2$

Rectangle
$A = lb$

Triangle
$A = \frac{1}{2}bh$

Parallelogram
$A = bh$

Trapezium
$A = \frac{1}{2}(a + b)h$

Circle
$A = \pi \dfrac{d^2}{4}$ or πr^2

EXAMPLES

1. An arch has a semicircular top as shown in Fig. 6.5. Calculate the area of the surrounding brickwork (i.e. the cross-section).

Figure 6.5 Arch with semicircular top

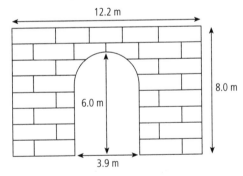

12.2 m

8.0 m

6.0 m

3.9 m

Take π as 3.14 and give your answer correct to 2 s.f.

We should first estimate the answer using the nearest whole numbers for the dimensions. The area of the brickwork is found by subtracting the area of the 'hole' from the whole figure.

Estimated area $= (12 \times 8) - [4 \times (6-2)] - \left(\dfrac{1}{2}\pi\dfrac{4^2}{4}\right) m^2$

$= 96 - 16 - 2\pi m^2$

$= 74\, m^2$ (taking π as 3)

Area A of brickwork required is

$$(12.2 \times 8.0) - \left[3.9 \times (6 - \frac{3.9}{2})\right] - \left(\frac{\pi \times 3.9 \times 3.9}{2 \times 4}\right) m^2$$
$$= (97.60 - 15.80 - 5.97)\,m^2 \text{ (taking } \pi \text{ as 3.14)}$$
$$= 75.83\,m^2$$
$$\Rightarrow A = 76\,m^2 \text{ (2 s.f.)}$$

Comparing this with the estimated value we see that the answer is sensible. Now work through this example yourself using a calculator.

2. The cross-section of a pipe is shown in Fig. 6.6. Find the area of the annulus (shaded portion) if D is 200 mm and d is 160 mm.

Figure 6.6 Cross-section of a pipe

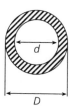

Using radius or diameter in the formula for area:

$$A = \pi R^2 - \pi r^2 \quad \text{OR} \quad \frac{\pi D^2}{4} - \frac{\pi d^2}{4}$$

$$A = \pi(R^2 - r^2)$$
$$= \pi(100^2 - 80^2)\,mm^2$$
$$= \pi(10\,000 - 6400)\,mm^2$$
$$= 100\pi(100 - 64)\,mm^2 \qquad \text{[taking out the common factor]}$$
$$\Rightarrow A = 3600\pi\,mm^2$$

We often leave answers in terms of π and leave a square root unworked unless the actual value is required. Work through this example yourself using the formula for diameters.

EXERCISE 6.2 Calculation of areas

Calculate the areas of the shaded shapes in Fig. 6.7. For numbers 4 and 6 give your answer in terms of π.

Surface areas of some three-dimensional objects

Some objects can be opened out into plane figures as shown in Fig. 6.8. (These plane figures are called the 'nets' of the objects.)

1. The **cube** of side l opens into six **squares** each of area l^2.

    Thus surface area of the cube $\quad A = 6l^2$

2. The curved surface of an **open cylinder** of diameter d and height h opens out into a **rectangle** of length πd and breadth h.

    Thus surface area of the cylinder $\quad A = \pi dh$

3. The curved surface of an **open cone** of slant height l and diameter d opens out into a sector of a circle of radius l and arc length πd.

Figure 6.7 Shapes for Exercise 6.2

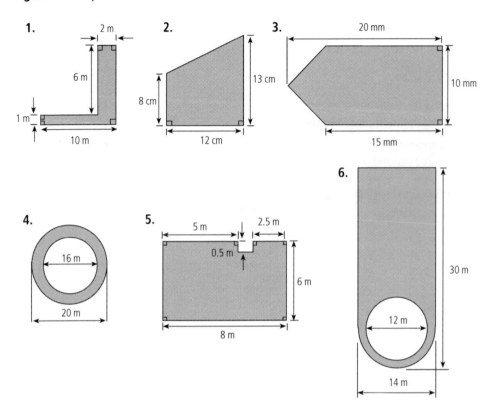

Figure 6.8 Nets of three-dimensional objects

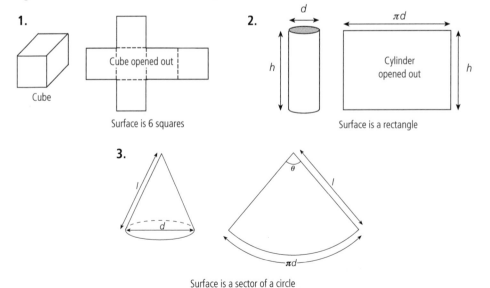

Thus surface area of cone $A = \dfrac{\theta°}{360°} \times \pi l^2$

$$\text{OR} \quad A = \dfrac{\pi d l}{2} \qquad \text{(proof not given)}$$

EXAMPLES

[NOTE: The following questions assume the ideal case of no waste.]

1. The surface area of a metal cube is 96 mm². What is the length of each side?
Total surface area A of a cube of side l is $6l^2$

$A = 6l^2$
$6l^2 = 96$
$l^2 = 16$ (dividing both sides by 6)
$l = \pm 4$
$l = 4\,\text{mm}$

We discard the negative square root, –4, because it is not a sensible answer.

2. Find the area of sheet metal required to make an open cone of diameter 80 mm and slant height 120 mm (see Fig. 6.9).

Figure 6.9 Open cone

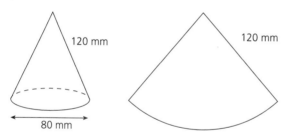

120 mm   120 mm

80 mm

We are given the diameter d and slant height l of the cone, so we use the formula $A = \pi \dfrac{d}{2} l$

Thus $A = \pi \times 40 \times 120\,\text{mm}^2$
$A = 4800\,\pi\,\text{mm}^2$

NOTE: if the cone were closed, then the area of the base, i.e. $\pi \times (40)^2\,\text{mm}^2$, would be added.

EXERCISE 6.3 Calculation of surface areas

The following questions assume no wasted metal. Make sketches for all questions. You do not need a calculator. Give answer in terms of π where appropriate.

1. Find the area of sheet metal required to make a cube of side
   (a) 30 mm
   (b) 20 mm

2. A closed metal cylinder has a height of 120 mm and a diameter of 50 mm. What area of sheet metal was required to make
   (a) the curved surface?
   (b) the total surface?

3. The area of sheet metal required to make an open cylinder of diameter 40 mm is 4800π mm². What will be
   (a) the height of the cylinder?
   (b) the additional area of metal required if the cylinder is to be closed?

4. A piece of metal in the shape of a sector of a circle is to be made into an open cone. What area of metal would be required if the slant height of the cone is to be 60 mm and the angle of the sector is 60°?

6.2 Volume

The **volume** of an object is a measure of the space it occupies.

Calculation of volume

Volume is measured by multiplying three lengths together. Units of volume are e.g. mm³, m³. We say that volume has three 'dimensions'. Figure 6.10 gives the formulae for volumes of some common shapes. Note that for solids with constant cross-section, volume = area of cross-section × length.

EXAMPLE

Rock is excavated from a hill in order to make a tunnel. The length of the tunnel is 400 m and the cross-section is as shown in Fig. 6.11.

1. Calculate the volume of rock removed.

2. Calculate the mass of rock removed.

   Density of rock is 2490 kg m⁻³ (to 3 s.f.).

   Take π as 3.14.

   Note carefully the context of the calculation and the number of significant figures used.

   The length of the tunnel is given to the nearest m, and the cross-sectional measurement to the nearest 10 cm. Do you see that this makes sense?

1. Volume of rock removed = area of cross-section × length.

$$V = 400[(7.00 \times 7.20) + \frac{1}{2} \frac{\pi(7.00)^2}{4}]\,m^3$$

$$= 400[50.4 + \frac{49.0\pi}{8}]\,m^3$$

$$= 400[50.4 + 19.2]\,m^3$$

$$\Rightarrow V = 27\,800\,m^3 \quad \text{to 3 s.f.}$$

Do you see why a more accurate answer would not be sensible?

2. Mass m = density ρ × volume V.

$$m = 2490 \times 400 \times 69.6\,kg$$

$$= 60\,321\,600\,kg$$

$$= 69\,300\,000\,kg \quad \text{(to 3 s.f.)}$$

$$\Rightarrow m = 6.93 \times 10^7\,kg \quad \text{(in standard form)}$$

Figure 6.10 Formulae for volume

Cuboid

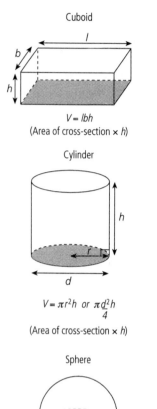

$V = lbh$
(Area of cross-section × h)

Cylinder

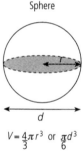

$V = \pi r^2 h$ or $\dfrac{\pi d^2 h}{4}$

(Area of cross-section × h)

Sphere

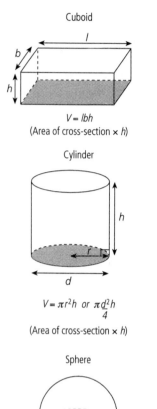

$V = \dfrac{4}{3}\pi r^3$ or $\dfrac{\pi}{6}d^3$

Triangular prism

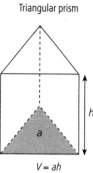

$V = ah$
(a is area of cross-section)

Cone

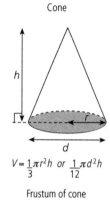

$V = \dfrac{1}{3}\pi r^2 h$ or $\dfrac{1}{12}\pi d^2 h$

Frustum of cone

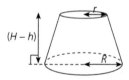

V = volume of whole cone – volume of small cone
$= \dfrac{1}{3}\pi R^2 H - \dfrac{1}{3}\pi r^2 h$
$= \dfrac{1}{3}\pi (R^2 H - r^2 h)$
(Note: H is height of whole cone not shown)
(h is height of small cone not shown)

Figure 6.11 Tunnel cross-section

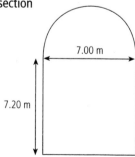

7.00 m

7.20 m

NOTE: (a) The numbers were not approximated until the last line.
(b) The large answer is best expressed in standard form.

Now work through the above example yourself.

EXERCISE 6.4 Calculation of volumes

1. to 5. Find the volume of each of the five solids in Fig. 6.12. (There is no need to use a
calculator for these questions.)

Figure 6.12 Shapes for Exercise 6.4

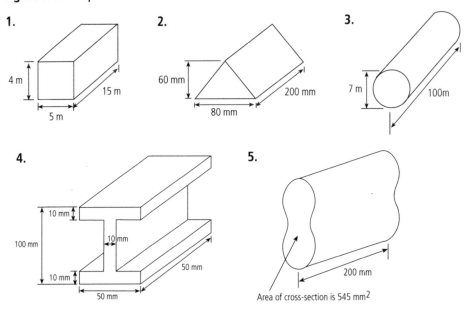

6. If the solid in number 4 is made of steel, calculate the mass of the solid given that the
density of steel is 7840 kg m^{-3} (to 3 s.f.). Give your answer in standard form.

[HINT: Remember to convert the volume from mm^3 to m^3]

7. The perpendicular height of a cone is 50 mm and the diameter is 60 mm. Make a sketch and
find the volume of the cone in terms of π.

8. A cone with a perpendicular height of 100 mm and a radius of 30 mm has a smaller cone
sliced off the top parallel to the base. This leaves a frustum with a perpendicular height of
60 mm and an upper circular face of radius 12 mm.

Draw a sketch and find the volume of the frustum in terms of π.

6.3 Areas and volumes of similar shapes

If a plane shape is enlarged or reduced, the two shapes are said to be **similar**
and all the corresponding sides will be in the same ratio. Similarity is
important in engineering and life, examples arising in construction and

Figure 6.13

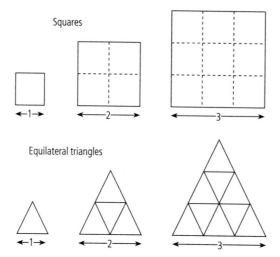

Figure 6.14

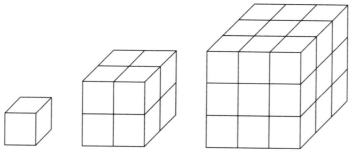

design, in maps, photography and even photocopying.

In Fig. 6.13 the ratio of corresponding sides for each set of shapes is 1:2:3; in the case of circles it will be the ratio of diameters. Similar shapes may be regular or irregular.

You will see that the corresponding ratio of areas is 1:4:9, i.e. $1^2:2^2:3^2$.

In Fig. 6.14 the corresponding ratio of volumes is 1:8:27, i.e. $1^3:2^3:3^3$.

Increasing or decreasing the linear ratio can thus have a dramatic effect on areas and volumes.

One example is the relationship between units:
$1\,m = 10^3\,mm$. Hence $1\,m^2 = (10^3)^2\,mm^2$ and $1\,m^3 = (10^3)^3\,mm^3$
$$= 10^6\,mm^2 \qquad\qquad = 10^9\,mm^3.$$

In general if $\dfrac{l}{L}$ is the linear scale factor (i.e. ratio)

$\left(\dfrac{l}{L}\right)^2$ is the area scale factor

$\left(\dfrac{l}{L}\right)^3$ is the volume scale factor.

EXAMPLES

1. A circle has a radius of 30 mm and is enlarged so that the radius is 45 mm. What is (a) the linear scale factor of enlargement, k? (b) the ratio of the enlarged area to the original area?

(a) $k = \dfrac{45}{30}$

$k = 1.5$

(b) $\dfrac{\text{enlarged area}}{\text{original area}} = (1.5)^2$

$= 2.25$

2. Two cylinders are similar in shape, their corresponding dimensions being in the ratio $1:3$. The volume of the larger cylinder V is 13 500 mm³. Calculate the volume of the smaller cylinder v.

Linear reduction factor $= \dfrac{1}{3}$

$$\therefore \frac{v}{V} = \left(\frac{1}{3}\right)^3$$

$$= \frac{1}{27}$$

$$\Rightarrow v = \frac{V}{27}$$

$$v = \frac{13\,500}{27}\,\text{mm}^3$$

$$v = 500\,\text{mm}^3$$

6.4 Lines and angles

In Fig. 6.15 P and Q are two points. A point marks a position and has no size. The line joining P and Q is called PQ and is the shortest path between these points. When two straight lines meet they form an angle.

Figure 6.15 Line PQ

P   Q

The angle between PQ and QR is written $\angle$PQR or PQ̂R. Angles are measured in degrees (°) and minutes (') (Fig. 6.16).

Figure 6.16 Angle PQR ($\angle$PQR or PQ̂R)

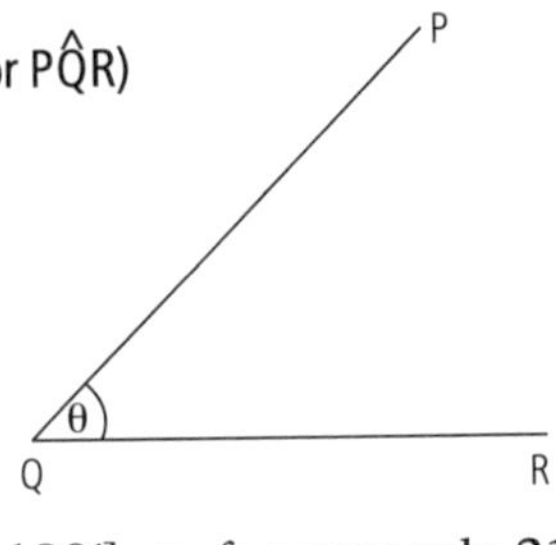

P

θ

Q   R

REMEMBER: $1° = 60'$ [NOT 100'], so for example 21°30' is 21.5° [NOT 21.3°].

Angles can also be measured in radians (rad) which we shall use later.

Figure 6.17 will remind you of the different types of angle and their sizes.

Figure 6.17 Types of angle

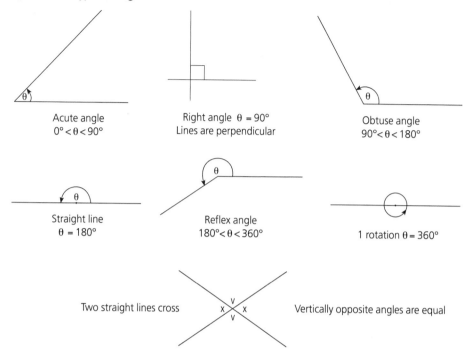

Acute angle
$0° < \theta < 90°$

Right angle $\theta = 90°$
Lines are perpendicular

Obtuse angle
$90° < \theta < 180°$

Straight line
$\theta = 180°$

Reflex angle
$180° < \theta < 360°$

1 rotation $\theta = 360°$

Two straight lines cross

Vertically opposite angles are equal

Angles are also formed with **parallel** lines. Parallel lines are lines which never meet however far they are extended. Corresponding angles (c) and alternate angles (a) are formed when a straight line crosses parallel lines (Fig. 6.18).

Figure 6.18 Parallel lines

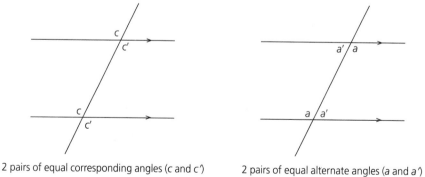

2 pairs of equal corresponding angles (c and c′)

2 pairs of equal alternate angles (a and a′)

REMEMBER: (a) Corresponding angles are equal

(b) Alternate angles are equal

Can you pick out the other pairs of corresponding angles in Fig. 6.18?

EXAMPLES

Figure 6.19 gives examples of finding unknown angles together with the reasons.

Figure 6.19 Finding unknown angles

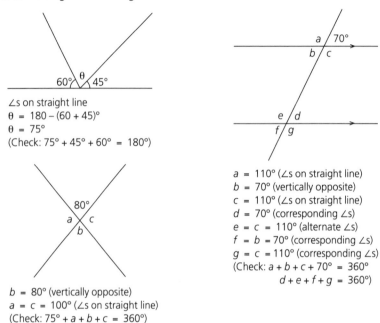

∠s on straight line
θ = 180 − (60 + 45)°
θ = 75°
(Check: 75° + 45° + 60° = 180°)

b = 80° (vertically opposite)
a = c = 100° (∠s on straight line)
(Check: 75° + a + b + c = 360°)

a = 110° (∠s on straight line)
b = 70° (vertically opposite)
c = 110° (∠s on straight line)
d = 70° (corresponding ∠s)
e = c = 110° (alternate ∠s)
f = b = 70° (corresponding ∠s)
g = c = 110° (corresponding ∠s)
(Check: a + b + c + 70° = 360°
d + e + f + g = 360°)

You will see that there is often more than one way to calculate the unknown angles.

EXERCISE 6.5 Finding angles made by straight lines

Calculate the unknown angles in Fig. 6.20, working quickly and giving reasons for your decisions.

6.5 Triangles and angles

A triangle has three sides. There are three types of triangle shown in Fig. 6.21. All triangles have one property in common, i.e. the three interior angles of a triangle total 180° (Fig. 6.22). That is why each angle of an equilateral triangle is 60°.

EXAMPLE

Calculate the unknown angles in Fig. 6.23.

EXERCISE 6.6 Finding angles made by parallel lines

Calculate the unknown angles in Fig. 6.24, giving reasons and working quickly.

Figure 6.20 Angles made by straight lines

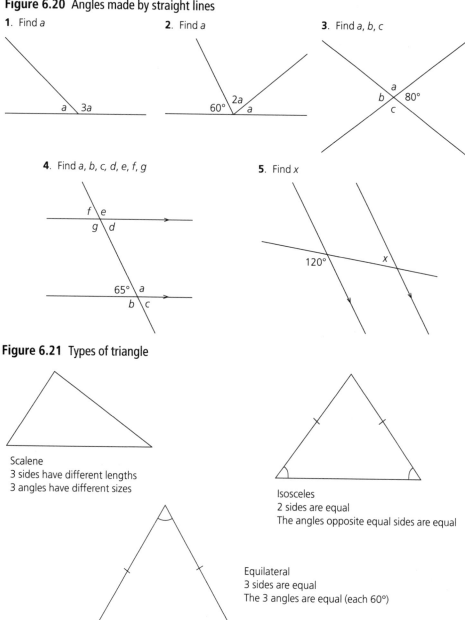

1. Find a

2. Find a

3. Find a, b, c

4. Find a, b, c, d, e, f, g

5. Find x

Figure 6.21 Types of triangle

Scalene
3 sides have different lengths
3 angles have different sizes

Isosceles
2 sides are equal
The angles opposite equal sides are equal

Equilateral
3 sides are equal
The 3 angles are equal (each 60°)

More properties of triangles

Triangles which are identical are called **congruent**. This means that all sides in the first triangle are equal to the corresponding sides in the second triangle. It also means that all angles in the first triangle are equal to the corresponding angles in the second triangle (see Fig. 6.25).

Two triangles which have the same shape but different size are called **similar triangles**. It follows that the angles in the first triangle are the same size as the angles in the second triangle (see Fig. 6.26).

Figure 6.22 Sum of interior angles of a triangle

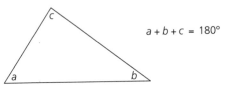

$$a + b + c = 180°$$

Figure 6.23 Finding unknown angles

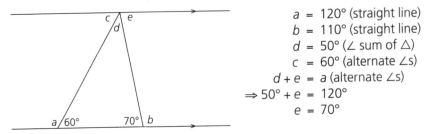

$$
\begin{aligned}
a &= 120° \text{ (straight line)} \\
b &= 110° \text{ (straight line)} \\
d &= 50° \text{ ($\angle$ sum of $\triangle$)} \\
c &= 60° \text{ (alternate $\angle$s)} \\
d + e &= a \text{ (alternate $\angle$s)} \\
\Rightarrow 50° + e &= 120° \\
e &= 70°
\end{aligned}
$$

Figure 6.24 Unknown angles

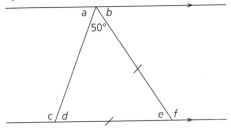

Figure 6.25   Congruent triangles

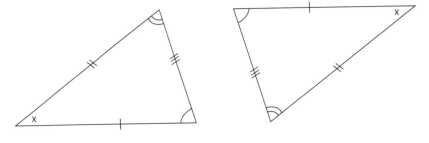

Figure 6.26   Similar triangles

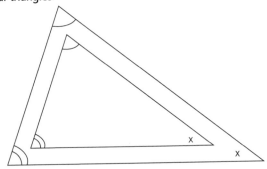

6.6 Quadrilaterals and angles

A quadrilateral has four sides. There are special types of quadrilateral which are shown in Fig. 6.27.

Figure 6.27 Quadrilaterals

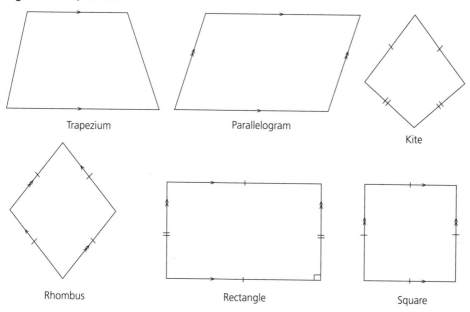

A line joining two opposite vertices of a quadrilateral is called a **diagonal**, as shown in Fig. 6.28.

Figure 6.28 Diagonal

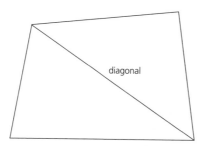

Can you see that: 'the sum of the angles of any quadrilateral is 360°'?

[HINT: The sum of the angles of each triangle is 180°.]

There are other properties which it is useful to know. These are shown in Fig. 6.29.

Figure 6.29 Properties of quadrilaterals

1. Parallelogram

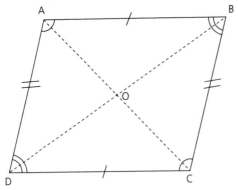

(i) AB = CD, AD = BC
(ii) ∠A = ∠C, ∠B = ∠D
(iii) AO = OC, BO = OD

2. Rhombus

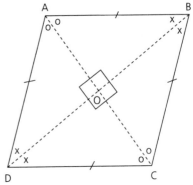

(i) Angles at O are right angles
(ii) ∠A = ∠C and each angle is bisected
     ∠B = ∠D and each angle is bisected

3. Rectangle

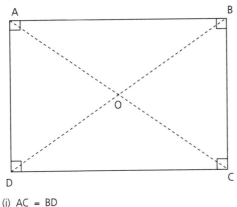

(i) AC = BD

4. Square

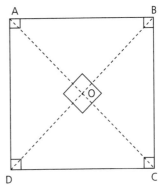

(i) Angles at O are right angles
All angles formed in the square ABCD
are either 90° or 45°

You see from Fig. 6.29 that:

1. all **parallelograms** have:
   (a) opposite sides equal
   (b) opposite angles equal
   (c) diagonals which bisect each other.

2. **rhombuses** have all the properties of a parallelogram. In addition:
   (a) diagonals bisect each other at 90°
   (b) diagonals bisect opposite angles.

3. **rectangles** are parallelograms with all angles of 90°.
   In addition:
   (a) the diagonals are equal.

4. **squares** are rectangles with all sides equal.
   In addition:
   (a) the diagonals meet at right angles.

EXAMPLE

In Fig. 6.30 PQRS is a rhombus.

Figure 6.30 Finding angles in a rhombus

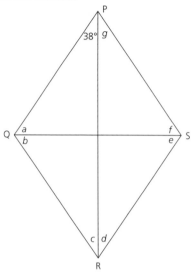

Calculate angles a to g, giving reasons.

PQ = QR = RS = PS ∴ △s PQR and PSR are isosceles. Also diagonals PR and QS bisect the angles
∴ 38° = c = d = g
Diagonals intersect at right angles
∴ a = 90°−38° = 52°
b = a = 52°
e = f = 52°

EXERCISE 6.7 Calculation of angles in a parallelogram

Calculate the unknown angles in Fig. 6.31, giving reasons and working quickly.

6.7 Circles and angles

The main properties of angles in circles are summarized in Fig. 6.32. Note that in a cyclic quadrilateral all four vertices lie on the circle.

EXAMPLE

Calculate the unknown angles in Fig. 6.33, giving reasons.
   Angles a and b are on the same arc as the angle of 86° at the centre.
$$\therefore a = b = \frac{1}{2} \times 86°$$
$$a = 43°$$
$$b = 43°$$

Figure 6.31 Calculation of angles in a parallelogram

1. ABCD is a parallelogram.   Find x, y, and c

2. LMNP is a rectangle. LQ = LP. Find a, b and c

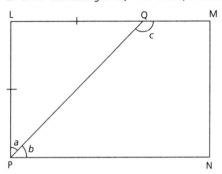

3. EFGH is a rhombus. ∠EFH is 40°. Find a, b and c

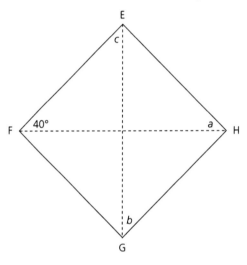

EXERCISE 6.8 Finding angles in a circle

Calculate the unknown angles in Fig. 6.34, giving reasons and working quickly.

You will find it very useful to have these basic ideas of geometry 'at your fingertips' because engineering is concerned with structures, their shapes, areas and space occupied.

6.8 Pythagoras's theorem

Pythagoras's theorem states that **all** right-angled triangles obey the following rule:

> The square on the hypotenuse is equal to the sum of the squares on the other two sides.

Figure 6.35 gives the general equation and the specific example of a 3, 4, 5 triangle.

Figure 6.32 Properties of angles in a circle

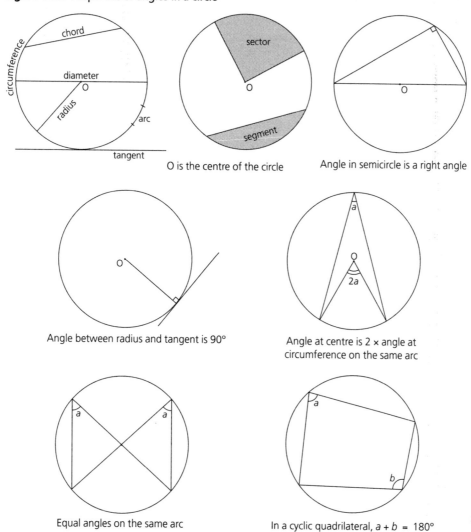

O is the centre of the circle

Angle in semicircle is a right angle

Angle between radius and tangent is 90°

Angle at centre is 2 × angle at circumference on the same arc

Equal angles on the same arc

In a cyclic quadrilateral, $a + b = 180°$

Figure 6.33 Example: finding unknown angles in a circle

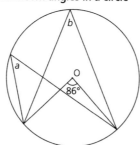

EXAMPLE

A garden shed is to be erected with a rectangular base 2.80 m long and 2.10 m wide (Fig. 6.36). One diagonal has to be marked out so as to ensure right angles at each corner. Calculate the length of this diagonal.

Figure 6.34 Finding angles in a circle

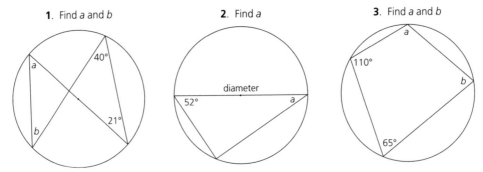

1. Find a and b 2. Find a 3. Find a and b

Figure 6.35 Pythagoras's theorem

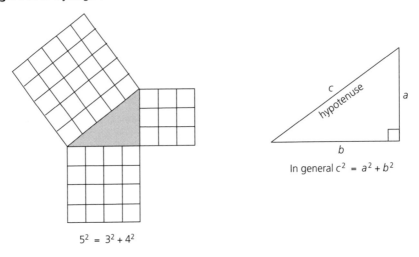

$$5^2 = 3^2 + 4^2$$

In general $c^2 = a^2 + b^2$

Figure 6.36 Base of a garden shed

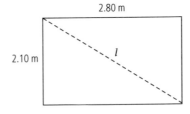

2.80 m

2.10 m

First we estimate the size of the answer. Rounding 2.10 to 2 and 2.80 to 3,

$$l^2 \approx 2^2 + 3^2$$
$$l \approx \sqrt{13}$$
$$\Rightarrow l \text{ must be between 3 and 4.}$$

Now we use the calculator.

$$l^2 = 2.10^2 + 2.80^2$$
$$= 4.41 + 7.84$$
$$l = \sqrt{12.25}$$
$$l = 3.50 \text{ m} \quad (3 \text{ s.f.})$$

NOTE: This calculation was carried out on one scientific calculator in the following sequence.

$l =$ ON/C | 2.10 | x^2 | + | 2.80 | x^2 | = | √ | =

$l = 3.50\,m$ (3 s.f.)

Your calculator may have a different sequence. Read the instructions carefully.

EXERCISE 6.9 Using Pythagoras's theorem

Use a calculator and give your answer to 3 s.f. Remember to estimate first.

1. Figure 6.37 shows a concrete post supported by two wire struts of equal length. Calculate the length of a strut.

Figure 6.37 Concrete post supported by wire struts

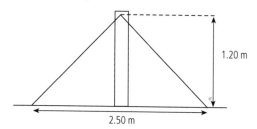

1.20 m

2.50 m

2. Figure 6.38 shows a side-view of a dormer window in a building. Calculate how far the window projects from the roof.

Figure 6.38 Dormer window

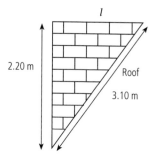

l

2.20 m

Roof

3.10 m

3. Calculate the value of h in Fig. 6.39.

Figure 6.39   **Figure 6.40**   **Figure 6.41**

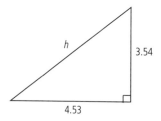
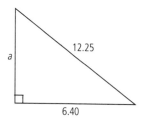
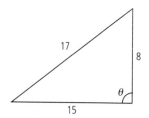

h

3.54

4.53

12.25

a

6.40

17

8

θ

15

4. Calculate the value of a in Fig. 6.40.

5. Show, by calculation, that θ is 90° (Fig. 6.41).   [HINT: Work out the squares on all sides.]

6.9 Angles measured in radians

Degrees and radians are the basic units in which angles are measured. Using degrees, one complete rotation of a radial line is defined as a rotation through 360°. Using radians, one complete rotation would be 2π radians because of the way in which 1 radian (Fig. 6.42) is defined:

Figure 6.42 Degrees and radians

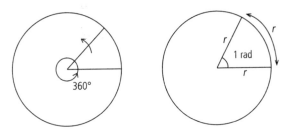

One radian is the angle subtended at the centre of a circle by an arc whose length is equal to the radius.

2π radians is the same angle as 360°

$\therefore$ 1 radian (rad) is the same angle as $\dfrac{360°}{2\pi}$, i.e. 57.3°

1 rad = 57.3°

In engineering, angles are often more often conveniently expressed in terms of π. Some useful ones to remember are given in Table 6.1.

Table 6.1 Equivalence between degrees and radians

Degrees	Radians
360	2π
180	π
90	$\dfrac{\pi}{2}$
45	$\dfrac{\pi}{4}$
60	$\dfrac{\pi}{3}$

EXAMPLES

Use a calculator.
1. Convert the following angles in degrees into radian measure.
    (a) 65°   (b) 306°

    (a) $65° = \dfrac{65}{360} \times 2\pi$ rad   (b) $306° = \dfrac{306}{360} \times 2\pi$ rad

    $65° = 1.13$ (2 d.p.) rad $306° = 5.34$ (2 d.p.) rad

REMEMBER: 1 rad is approx. 60°. Use this approximation to see if the answer makes sense, e.g. 306° should be slightly more than 5 rad.

2. Convert the angles in question 1 into radian measure in terms of π.

(a) $65° = \dfrac{65}{360} \times 2\pi$ rad

$65° = 0.36\,\pi$ rad

(b) $306° = \dfrac{306}{360} \times 2\pi$ rad

$306° = 1.7\,\pi$ rad

3. Convert the following angles in radians to degree measure.

(a) 0.9 rad

(b) 2.1 rad

(a) $0.9\,\text{rad} = \dfrac{0.9}{2\pi} \times 360°$

$0.9\,\text{rad} = 51.6°$ (1 d.p.)

(b) $2.1\,\text{rad} = \dfrac{2.1}{2\pi} \times 360°$

$2.1\,\text{rad} = 120.3°$ (1 d.p.)

Use the same approximation, i.e. 1 rad ≈ 60°, to check your answer.

EXERCISE 6.10 Converting degrees and radians

Copy Tables 6.2 and 6.3 and fill in the missing values. Use a calculator only for Table 6.3.

Table 6.2 Equivalence of degrees and radians (with π)

Degrees	Radians with π
20	$\dfrac{\pi}{9}$
60	
80	
100	
120	
240	
300	
360	

Table 6.3 Equivalence of degrees and radians

Degrees (nearest °)	Radians (2 d.p.)
20	0.35
60	
120	
240	
300	
	0.52
	1.40
	1.75
	3.49
	6.28

6.10 Trigonometric ratios

You can see in Fig. 6.43 that the right-angled triangles with the same angle θ are similar, i.e. ratios of corresponding sides are the same.

Figure 6.43 Set of similar right-angled triangles

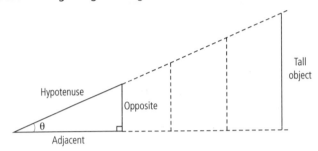

Thus each of the ratios opposite:hypotenuse, adjacent:hypotenuse, opposite:adjacent, has some constant value. These ratios between the sides occur frequently in maths and engineering. They are called **trigonometric ratios** and are defined as follows:

1. sine of angle θ, written as sin θ.

$$\sin \theta = \frac{\text{opposite}}{\text{hypotenuse}}$$

2. cosine of angle θ, written as cos θ.

$$\cos \theta = \frac{\text{adjacent}}{\text{hypotenuse}}$$

3. tangent of angle θ, written as tan θ.

$$\tan \theta = \frac{\text{opposite}}{\text{adjacent}}$$

If you look at these ratios carefully, you will see that

$$\tan \theta = \frac{\sin \theta}{\cos \theta}$$

Each of these ratios has the same value for a given θ.

Look carefully at Fig. 6.44. Now apply the ratios to these angles and see how Table 6.4 is formed.

Look carefully at Table 6.4. You will observe that, for $0 \leq \theta \leq 90°$ (i.e. θ having values between and including 0° to 90°),

1. $0 \leq \sin \theta \leq 1$, i.e. values of sin θ increase from 0 to 1
2. $1 \leq \cos \theta \leq 0$, i.e. values of cos θ decrease from 1 to 0
3. $0 \leq \tan \theta \leq \infty$, i.e. values of tan θ increase from 0 to ≈ 57 when θ is 89° to infinity when θ is 90°.

Figure 6.44 Triangle facts

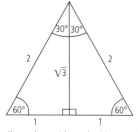

Equilateral △ with each side 2 units
(use Pythagoras to find √3)

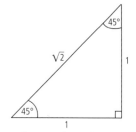

Isosceles △ with equal sides 1 unit
(use Pythagoras to find √2)

Table 6.4 Values of sin θ, cos θ and tan θ

θ (rad)	θ (°)	sin θ	cos θ	tan θ
0	0	0	1	0
$\frac{\pi}{6}$	30	$\frac{1}{2}$	$\frac{\sqrt{3}}{2}$	$\frac{1}{\sqrt{3}}$
$\frac{\pi}{3}$	60	$\frac{\sqrt{3}}{2}$	$\frac{1}{2}$	$\sqrt{3}$
$\frac{\pi}{4}$	45	$\frac{1}{\sqrt{2}}$	$\frac{1}{\sqrt{2}}$	1
$\frac{\pi}{2}$	90	1	0	∞*

*Symbol means 'infinity'

REMEMBER: $\tan \theta = \dfrac{\sin \theta}{\cos \theta}$

$$\tan 90° = \dfrac{\sin 90°}{\cos 90°}$$

$$= \dfrac{1}{0}$$

$\Rightarrow \tan 90° \rightarrow$ infinity

[NOTE: This means that tan 90° is so large that it cannot be given any value.]

In Exercise 6.11 you will be asked to find values of the trigonometric ratios and sketch some curves, so that you can visualize the effects of increasing the value of θ.

You will find that memorizing the ratios in Table 6.4 can be very useful in your engineering work. Tables have been constructed of the ratios for all angles between 0° and 90° but of course the values inbuilt into a scientific calculator are much quicker to use.

Using a calculator to find trigonometric ratios

At this stage it would be helpful for you to make sure that you can use the ratios correctly on your calculator and in doing this you will build up knowledge of the way in which these ratios increase or decrease with increasing values of θ. It is important that you **read carefully the instruction booklet provided with your calculator.**

EXAMPLE

This example illustrates the use of one type of scientific calculator.

- The $\boxed{\text{DRG}}$ button specifies three angular units which change in the sequence DEG→RAD→GRAD. We are concerned only with DEG(°) and RAD (rad)

- Find sin 40°
  $\boxed{\text{ON/C}}$ $\boxed{\text{DRG}}$ (for DEG) $\boxed{\text{sin}}$ $\boxed{40}$ $\boxed{=}$ 0.6427 . . .

- Find cos $\frac{\pi}{4}$ (rad)
  $\boxed{\text{ON/C}}$ $\boxed{\text{DRG}}$ (for RAD) $\boxed{\text{cos}}$ $\boxed{\ }$ $\boxed{(}$ $\boxed{\text{2nd F}}$ $\boxed{\pi}$ $\boxed{\div}$ $\boxed{4}$ $\boxed{=}$ 0.7071 . . .

Again it is emphasized that these sequences of operation apply to one type of scientific calculator. Your calculator may operate differently.

EXERCISE 6.11 Finding trigonometric ratios

1. Copy Table 6.5 and use your calculator to find the values of sin θ and cos θ correct to 2 d.p. Look carefully at the completed table. What do you observe?

Table 6.5 Values for sin θ and cos θ

θ°	sin θ	cos θ
0		
10		
30		
45		
60		
80		
90		

2. Using the values you have obtained for Table 6.5 **sketch** a curve of sin θ with θ and a curve of cos θ with θ, using the same pair of axes. Plot θ along the horizontal axis and sin θ and cos θ along the vertical axis.

Note the relationship between sin θ and cos θ as θ increases from 0° to 90°.

3. Copy Table 6.6 and use your calculator to find the values of tan θ. For 0 ≤ θ ≤ 45° give the values correct to 2 d.p. For 45 ≤ θ ≤ 90° give the values to the nearest whole number.

Study the completed table carefully. What do you observe?

Table 6.6 Values for tan θ

θ°	tan θ
0	
10	
20	
30	
40	
45	
60	
70	
80	
85	
89	
89.5	
89.9	
90	

6.11 The inverse operations

Sometimes we are given the trigonometric ratio and are required to find the angle. For example consider the much used ratio sin θ = 0.5. We rearrange this as θ = sin⁻¹ 0.5 and using the sin⁻¹ operation on the calculator we get, as expected, θ = 30° (or 0.52 rad to 2 d.p.). The symbol sin⁻¹ on the calculator key indicates that we are carrying out the inverse operation from finding sin θ. In the same way the cos⁻¹, tan⁻¹ keys carry out the inverse operations from finding cos θ and tan θ.

[A strong note of caution! The inverse symbols do **not** mean the reciprocals $\frac{1}{\sin\theta}$, $\frac{1}{\cos\theta}$ or $\frac{1}{\tan\theta}$.]

EXAMPLE

Find θ in degrees and radians given that sin θ = 0.5.

Follow this example using your calculator according to the instructions. On one scientific calculator the procedure for finding sin⁻¹ 0.5 would be

ON/C 2nd F sin⁻¹ 0.5 = 30°
DRG→ 0.524 rad (3 d.p.)

but your calculator may operate differently.

θ = 30°   OR   0.524 rad (3 d.p.)

EXERCISE 6.12 Finding the angle given the trigonometric ratio

Use a calculator to find the following angles:
(a) in degrees to 1 d.p.   (b) in rad to 3 d.p.

1. $\sin^{-1} 0$, $\sin^{-1} 0.55$, $\sin^{-1} 0.8$, $\sin^{-1} 1$

2. $\cos^{-1} 0$, $\cos^{-1} 0.3$, $\cos^{-1} 0.9$, $\cos^{-1} 1$

3. $\tan^{-1} 0$, $\tan^{-1} 0.99$, $\tan^{-1} 0.5$, $\tan^{-1} 0$.

NOTE: You will sometimes meet angles measured in degrees and **minutes** rather than the more usual decimal form, e.g. 31°30′ rather than 31.5°. For calculations you will need to change to the decimal form. This is easily done provided you REMEMBER that **1° = 60′** (not 100′).

EXAMPLES

Change to decimal form:
1. 14°35′   2. 53°14′

Using the calculator to change the minutes to a decimal fraction we have:

1. $35' = \dfrac{35°}{60}$

   $= 0.6°$ (to 1 d.p.)

   The angle is 14.6°

2. $14' = \dfrac{14°}{60}$

   $= 0.2°$ (to 1 d.p.)

   The angle is 53.2°

Make up some practice examples for yourself.

EXAMPLES

1. Using Fig. 6.45

    (a) Write down the values of sin θ, cos θ, tan θ
    (b) Find the value of θ in degrees

Figure 6.45

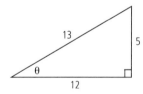

(a) $\sin \theta = \dfrac{5}{13}$, $\cos \theta = \dfrac{12}{13}$, $\tan \theta = \dfrac{5}{13}$

(b) θ could be found from any one of these ratios.

Taking $\sin \theta = \dfrac{5}{13}$ then $\theta = \sin^{-1} \dfrac{5}{13}$

$\theta = \sin^{-1} 0.38$

$\theta = 22.6°$

2. A guy wire of length 4.50 m to a pole rising from level ground makes an angle of 58° with the ground (Fig. 6.46). How high above the ground is the wire attached to the pole?

The unknown height h is the side opposite to ∠58°. We therefore use the ratio sin 58°.

$\sin 58° = \dfrac{h}{4.50}$

$h = 4.50 \times \sin 58°$

$h = 3.80 \, m$ (to 2 s.f.)

Figure 6.46

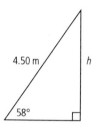

3. A coastguard watching a boat at sea sees the boat at an angle of depression of 10°. The coastguard is 75 m above sea level. How far (at sea level) is the boat from the coastguard?

The angle of depression 10° is equal to the alternate angle in the right-angled triangle and d is the adjacent side. We therefore use tan 10° OR tan 80° (which puts d 'on top') (see Fig. 6.47)

Figure 6.47

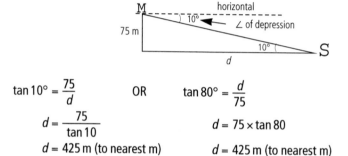

$$\tan 10° = \frac{75}{d} \qquad \text{OR} \qquad \tan 80° = \frac{d}{75}$$

$$d = \frac{75}{\tan 10} \qquad\qquad d = 75 \times \tan 80$$

$$d = 425 \text{ m (to nearest m)} \qquad d = 425 \text{ m (to nearest m)}$$

NOTE: Angles of elevation and depression are always measured from the horizontal with the line of sight. In this problem a person on the boat would see the coastguard at an **angle of elevation** of 10°.

4. P and Q are two villages 25 km apart. If Q is on a bearing 55° east of north from P, how far east is Q from P? (See Fig. 6.48.)

Figure 6.48

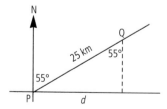

The unknown distance d is side opposite to the alternate ∠55° in the right-angled triangle. We therefore use sin 55°.

$$\sin 55° = \frac{d}{25}$$

$$d = 25 \times \sin 55°$$

$$d = 20 \text{ km (to nearest km)}$$

NOTE: To find the bearing of a point Q from a point P first join P and Q. Then draw in the north line at P and find the angle between north and PQ measured **clockwise**.

5. In a reciprocating engine, shown in Fig. 6.49, the crank is 10 cm long and the connecting rod is 30 cm long. Find ∠θ.

Figure 6.49

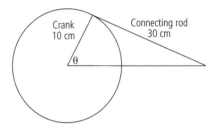

Assume that the connecting rod is tangential to the circle and therefore at a right angle to the crank. We are given the sides opposite and adjacent to the unknown angle θ, therefore we shall use tan θ.

$$\tan \theta = \frac{30}{10}$$
$$\theta = \tan^{-1} 3$$
$$\theta = 71.6°$$

You should now be ready to tackle some problems yourself.

EXERCISE 6.13 Trigonometric problems

In each of the first four questions it is essential that you make a sketch and put in the information.

1. In a 3, 4, 5 triangle where θ is the angle between sides 3 and 5:
   (a) Write down the values of sin θ, cos θ, tan θ.
   (b) Find the value of θ in degrees (1 d.p.).

2. A ladder of length 5.80 m leans against a vertical wall with its bottom on horizontal ground. If the ladder makes an angle of 65° with the ground, how far is the bottom end from the wall? (2 s.f.)

3. A ramp is to be built with a slope of 12° so that a wheelbarrow can be wheeled up to a height of 1.50 m. Find the length of ramp required (2 s.f.).

4. A roof has one slope pitched at 35° to the horizontal and the other at 55°. If the height of the eaves is 6.0 m what is the span of the roof?
   [HINT: There are two right-angled triangles.]

5. In Fig. 6.50 find the value of h and ∠θ.

Figure 6.50

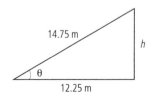

14.75 m

h

θ

12.25 m

6.12 Adding scalars and vectors

Scalars are specified by magnitude only, e.g. mass. Scalars can be simply added or subtracted. A mass of 4 kg added to a mass of 3 kg gives a mass of 7 kg.

Vectors must be specified in magnitude and direction, e.g. force, velocity. Vectors in the same direction can be added, but what if the vectors are at an angle? Common sense suggests that the effect of two forces, say, acting on a body, is that of one larger force acting in a direction between them. The parallelogram rule enables us to add vectors and find the one effective force.

The **parallelogram rule** states that:

> If two vectors **P** and **Q** are represented in size and direction by the two adjacent sides of a parallelogram, the diagonal represents the resultant in size and direction.

The resultant can be found from a scale drawing or by calculation. At this stage of your studies the calculation would be confined to the special case where **P** and **Q** are at right angles, as in Fig. 6.51. The resultant is then calculated using Pythagoras's theorem and its direction found from tan θ.

Figure 6.51 The parallelogram of forces

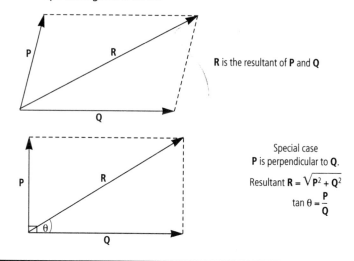

R is the resultant of **P** and **Q**

Special case
P is perpendicular to **Q**.
Resultant $R = \sqrt{P^2 + Q^2}$
$$\tan \theta = \frac{P}{Q}$$

EXAMPLE

Forces of 4 N and 3 N act together at a point. Find the resultant force when they are:
(a) in the same line and in the same direction
(b) in the same line but in opposite directions
(c) perpendicular to each other

Draw sketches to illustrate the answers.

(a)   4 N →
      3 N →   7 N →

The resultant is 7 N in the original direction.

(b) $\xleftarrow{\quad 3\,N\quad}\;\Big|\;\xrightarrow{\quad 4\,N\quad}$ $\qquad\qquad\xrightarrow{1\,N}$

The resultant is 1 N in the direction of the 4 N force.

(c) **R** is the resultant

$$R = 5\,N \text{ using Pythagoras}$$

$$\tan\theta = \frac{3}{4} \quad \text{so} \quad \theta = 37° \text{ (to nearest degree)}$$

The resultant is 5 N in a direction at 37° to the 4 N force (see Fig. 6.52).

Figure 6.52

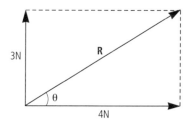

EXERCISE 6.14 The parallelogram rule

Use trigonometry to solve the problems. Remember that sketches are essential.

1. Forces of 5 N and 12 N act together at a point. Find the resultant force when they are
    (a) in the same line and in the same direction.
    (b) in the same line but in opposite directions
    (c) perpendicular to each other.

2. Repeat question 1 if the forces are 10.5 N and 7.5 N.

3. In a plane structure a point is acted on by forces of 1.5 N and 2.0 kN in the plane. If the forces are at right angles, what is the magnitude and direction of the resultant force?

4. Find the resultant velocity in magnitude and direction of velocities of 5.0 km/h due north and 8.0 km/h due east.
    Show on a sketch diagram the direction of the resultant velocity.

5. A river is 100 m wide and is flowing at 1.5 km/h. Find:
    (a) the effective velocity of a swimmer who can swim at 2 km/h in still water
    (b) the time taken to swim across the river.

6.13 Resolving vectors

We have seen how two vectors can be added to give a single vector. There is a reverse process of taking a single vector and expressing it in terms of two component vectors.

In Fig. 6.53 you see that a single vector **P** can be resolved into two components c_1 and c_2 at right angles.

Figure 6.53 Resolution of a vector

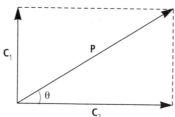

Using trigonometry $\sin \theta = \dfrac{c_1}{P}$

$$\cos \theta = \dfrac{c_2}{P}$$

$$\Rightarrow c_1 = P \sin \theta$$
$$\text{and } c_2 = P \cos \theta$$

This technique is particularly useful in problems on motion and forces.

EXAMPLES

1. Find the horizontal and vertical components of an 80.0 N force acting at an angle of 60° to
    the horizontal (see Fig. 6.54).
    Horizontal component = 80.0 cos 60°
    $\qquad\qquad\qquad\quad$ = 40.0 N
    Vertical component = 80.0 sin 60°
    $\qquad\qquad\qquad\;$ = 69.3 N

Figure 6.54

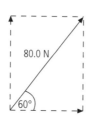

Figure 6.55

2. A projectile has an initial velocity of 5.0 m/s at an angle of 45° to the horizontal. What are
    the horizontal and vertical components of this velocity? (See Fig. 6.55.)
    Horizontal component = 5.0 cos 45° m/s
    $\qquad\qquad\qquad\quad$ = 3.5 m/s
    Vertical component = 5.0 sin 45° m/s
    $\qquad\qquad\qquad\;$ = 3.5 m/s

3. An object of weight 20.0 N rests on an incline at 30° to the horizontal. Find the components
    of the weight acting at right angles to the plane and along the incline (see Fig. 6.56).

Figure 6.56

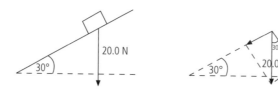

$$\text{Component along the incline} = 20.0 \sin 30° \, \text{N}$$
$$= 10.0 \, \text{N}$$
$$\text{Component at right angles to the incline} = 20.0 \cos 30° \, \text{N}$$
$$= 17.3 \, \text{N}$$

EXERCISE 6.15 Resolving vectors

Make sketches for all the following questions.

1. Find the horizontal and vertical components of the forces:
   (a) 25.0 N at 30° to the horizontal
   (b) 20.0 N at 50° to the horizontal
   (c) 14.0 N at 35° to the horizontal
   (d) 10.0 kN at 30° to the vertical
   (e) 10.0 kN at 45° to the vertical.

2. An object of weight 28.0 N rests on an incline which is at 15° to the horizontal. Find the components of the weight at right angles to the incline and parallel to the incline.

3. Repeat question 2 for a weight of 20.0 N and an angle of 12°.

4. A projectile has a velocity of 100 m/s in a north-easterly direction. Find the components of this velocity in the north, east, south and west directions. (Think carefully)!

5. Find the horizontal and vertical components of the velocities:
   (a) 200 m/s at 30° to the horizontal
   (b) 200 m/s at 60° to the horizontal
   (c) 500 m/s at 45° to the horizontal.

6. A force of 10.0 N acts at an angle of 45° to the horizontal. Another force of 15.0 N acting at the same point is at 60° to the horizontal. Find the sum of:
   (a) the horizontal components
   (b) the vertical components.

6.14 Introduction to phasors

Phasors (or phase vectors) can be used to represent alternating currents and voltages. The magnitude of a phasor can represent either the root mean square (r.m.s.) value or the peak value of the alternating quantity, provided we are consistent.

The phasor diagram in Fig. 6.57 shows two voltages of the same frequency which differ in phase by one quarter of a cycle, i.e. by $\frac{\pi}{2}$ or 90°. By convention,

Figure 6.57 Phasor diagram

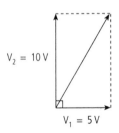

$V_2 = 10 \, \text{V}$

$V_1 = 5 \, \text{V}$

anticlockwise differences in angle mean leading in phase. Thus voltage V_2 leads voltage V_1 by $\pi/2$.

The same parallelogram law can be used for adding phasors as for adding vectors. Thus the resultant voltage when phasors V_1 and V_2 are added is

$$V = \sqrt{V_1^2 + V_2^2}$$

and V leads V_1 by an angle ϕ where

$$\tan \phi = \frac{V_2}{V_1}$$

The phasors must both be voltages or both be currents. Also they must be of the same frequency if they are to be added by this method.

Phasor diagrams can also be used to show the phase difference between current and voltage in a component or circuit carrying a.c. For example,

- in a pure capacitor the current leads the voltage by 90°
- in a pure inductor the voltage leads the current by 90°
- in a pure resistor the voltage and the current are in phase.

These are illustrated in Fig. 6.58.

Figure 6.58 Phase difference in an a.c. circuit

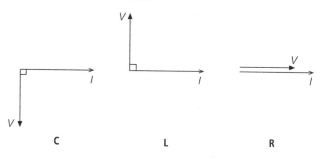

EXERCISE 6.16 Phasors

1. Figure 6.57 shows a phasor diagram drawn using r.m.s. values for two alternating voltages of the same frequency.
   (a) Which voltage leads in phase?
   (b) What is the phase difference between V_2 and V_1 in radians and in degrees?
   (c) What is the r.m.s. value $V_{r.m.s.}$ of the resultant voltage?
   (d) What is the peak value of the resultant voltage?

2. Figure 6.59 shows the voltage V and current I phasors for components in an a.c. circuit. In each case 1 to 6 what is the phase relationship between voltage and current in (a) degrees and (b) in radians?
   Which case is for (c) a capacitor, (d) an inductor, (e) a resistor?
   The voltages across two components in series are represented in case 6.
   (f) Find the r.m.s. value of the resultant voltage.
   (g) What is its peak value?
   (h) What is the phase relationship between the resultant voltage and the current?

Figure 6.59 Voltage and current phasors for components in an a.c. circuit

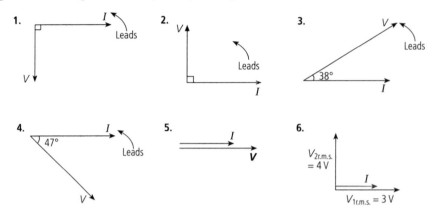

Chapter review

Read carefully through the summary. Many of the facts here will be useful when you are studying other chapters.

Summary

REMEMBER: When using formulae

- **Perimeter** has 1 dimension. Unit is mm, m.
  e.g. $P = 2l + 2b$ (rectangle), $c = \pi d$ (circle), π is a number
- **Area** has 2 dimensions. Unit is mm², m².
  e.g. $A = lb$ (rectangle), $A = \dfrac{\pi}{4} d^2$ (circle)
- **Volume** has 3 dimensions. Unit is mm³, m³.
  $V = lbh$ (cuboid), $V = \dfrac{1}{6}\pi d^3$ (sphere)
- For **similar** figures: $\dfrac{l}{L}$ is the linear scale factor

  $\dfrac{a}{A} = \left(\dfrac{l}{L}\right)^2$ is the area scale factor

  $\dfrac{v}{V} = \left(\dfrac{l}{L}\right)^3$ is the volume scale factor

REMEMBER: When converting units
  $1\,\text{m} = 10^3\,\text{mm}$; $1\,\text{m}^2 = 10^6\,\text{mm}^2$; $1\,\text{m}^3 = 10^9\,\text{mm}^3$

- **Angles**
  There are many different types of angle:
  acute $0 < \theta < 90°$,   obtuse $90 < \theta < 180°$,   reflex $180 < \theta < 360°$

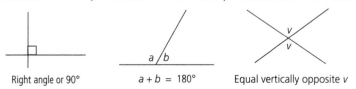

Right angle or 90° $a + b = 180°$ Equal vertically opposite v

- **Parallel lines**

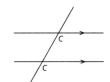

Equal corresponding angles c

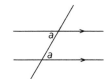

Equal alternate angles a

- **Triangles**

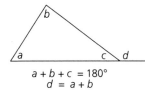

$a + b + c = 180°$
$d = a + b$

Equilateral triangle

Isosceles triangle

- **Quadrilaterals**

Parallelogram
Diagonals bisect

Rhombus
Diagonals bisect

Rectangle
Diagonals equal

Square
Angles 90° or 45°

- **Circles**

Tangent

$a + b + c + d = 180°$

- **Pythagoras's theorem**
$c^2 = a^2 + b^2$

Pythagoras

- **Angles** in degrees (°), OR radians (rad), 1 rad ≈ 60°
$360° = 2\pi$ rad, $180° = \pi$ rad

$$90° = \frac{\pi}{2} \text{rad}, \quad 45° = \frac{\pi}{4} \text{rad}$$

1 radian

- **Trigonometric ratios**

$$\sin \theta = \frac{\text{opp}}{\text{hyp}}$$

$$\cos \theta = \frac{\text{adj}}{\text{hyp}}$$

$$\tan \theta = \frac{\text{opp}}{\text{adj}} = \frac{\sin \theta}{\cos \theta}$$

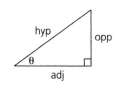

- **Memorize:**

  $0 \le \theta \le 90°$ (or $\frac{\pi}{2}$ rad), $0 \le \sin \theta \le 1$; $1 \le \cos \theta \le 0$; $0 \le \tan \theta \le \infty$

- **Inverse operations:** $\sin^{-1}$, $\cos^{-1}$, $\tan^{-1}$,
  i.e. find θ given the ratio

  [CAUTION: **Not** $\dfrac{1}{\sin}$, $\dfrac{1}{\cos}$, $\dfrac{1}{\tan}$.]

- **Scalar quantities**, e.g. mass, can be simply added or subtracted.

- **Vector quantities**, e.g. force, velocity, can be 'added' by the **parallelogram rule**. Resolving vectors into two components at right angles is the reverse process.

- **Phasors** (or phase vectors) are used to represent alternating currents and voltages. They are 'added' using the same parallelogram rule as for vectors.

chapter

7 Graphs

7.1 Introduction to line graphs

This chapter concentrates on 'line' graphs which may be **straight lines** or **curves**. They show how two quantities, *A* and *B*, are related. Figure 7.1a–f shows six ways in which line graphs are used:

1. displaying a picture of what happens to quantity *A* when quantity *B* changes (Fig. 7.1a)
2. plotting points to see if there is a relationship between *A* and *B* (Fig. 7.1b)

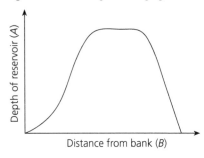

Figure 7.1a Showing '*A*' changing with '*B*'

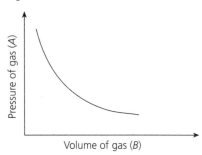

Figure 7.1b Is '*A*' related to '*B*'?

3. deducing the form of the relationship between *A* and *B* (Fig. 7.1c)
4. reading off values of *A* for given values of *B* (or vice versa) (Fig. 7.1d)

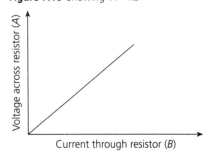

Figure 7.1c Showing '*A* = k*B*'

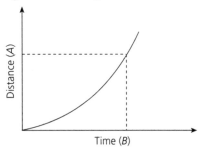

Figure 7.1d Reading off values

5. converting measurements in unit A to measurements in unit B (or vice versa) (Fig. 7.1e)
6. averaging experimental results and using the graph to make calculations (Fig. 7.1f).

Figure 7.1e Converting unit A to unit B

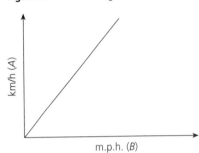

Figure 7.1f Result of Young modulus experiment

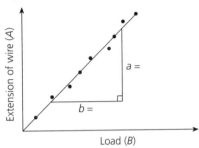

In the chapter on statistics (Chapter 4) you have met graphs of other forms.

The graph paper you use is generally divided into small squares of side 1 mm or 2 mm. When you plot a graph of A against B, A is chosen to be the dependent variable (the one you measure or calculate) and B is the independent variable (the one you set). You need to:

1. Select a title.
2. Use a **sharp** pencil.
3. Choose scales for the A axis and B axis.
4. Rule the axes with A vertical and B horizontal, using your graph paper to give the best possible display.
5. Mark the scale grids nearly on the axes.
6. Label each axis with quantity, symbol and unit.
7. Plot each point; mark it by ⊙ or × neatly. [NOT • or +]
8. Draw the graph as a neat line or curve.
9. If the graph is showing experimental results
    (a) make a rough graph as you go along
    (b) draw the line or curve 'of best fit'.

The graph in Fig. 7.2 illustrates that advice.

Figure 7.2 Graph of resistance against temperature for a copper coil

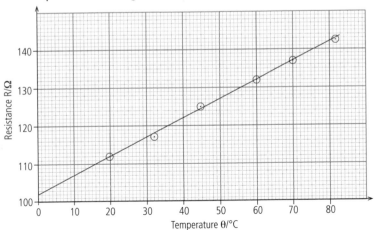

Figure 7.3 Lines of best fit

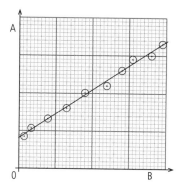 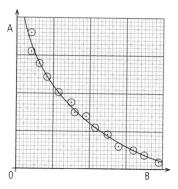

The line or curve you draw need not pass through every point but must have the points evenly distributed on both sides (see Fig. 7.3).

EXERCISE 7.1 Plotting a graph and reading off values

1. The resistance of a filament lamp is measured as the voltage across it is increased in 2 V steps from zero to just above its working voltage of 12 V. The results were

    0.5 Ω   1.5 Ω   2.2 Ω   3.2 Ω   4.2 Ω   5.0 Ω   6.0 Ω and 6.9 Ω

    Construct a table of voltage and resistance values for voltages from 0 to 14 V. Then plot a graph entitled 'Resistance against voltage for a filament lamp'. Label the vertical axis 'resistance R/Ω' and the horizontal axis 'voltage V/V'. Use a scale of 2 cm ≡ 1 Ω on the resistance axis and 1 cm ≡ 1 V on the voltage axis. Draw the line of best fit through the points.

2. Figure 7.4 gives the volume of gas coming from a cylinder, relating it to the gauge pressure for two different gases.
    (a) What volume of oxygen is available when the gauge reads 80 bars?
    (b) What volume of hydrogen is obtained when the gauge reads 170 bars?

Summary

Line graph: line or curve of A against B: A vertical, B horizontal.
REMEMBER: Title, suitable scales, grids marked, axes labelled (quantity, symbol, unit), sharp pencil, ⊙ or × for points, line of best fit, points evenly distributed.

7.2   Cartesian coordinates

We can locate any point in a plane using a system invented by Descartes – hence the name 'Cartesian'. x and y axes are drawn at right angles and each represents a number line passing through zero. The point where both x and y are zero is called the **origin**.

Any point is represented by **coordinates** (x, y) and is located by finding the

Figure 7.4 Volume of gas at 15°C and 1.013 bar from a gas cylinder

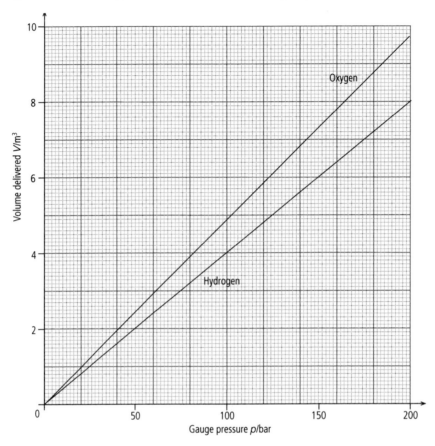

point with those values of x and y. The plane and the x and y axes extend
infinitely but when we need to plot points we select the area that we need.
For example, in Fig. 7.5, P is the point where $x = 3$ and $y = 2$, so we call it the
point (3,2).

Summary

y and x axes are drawn at right angles.
A point in the plane is represented by coordinates (x,y).

EXAMPLES

The following points have been plotted on the graph in Fig. 7.6.

A (4,4)   C (−2,4)   E (0, −1)
B (− 4,−4) D (− 1,−4) F (3.5,−2.5)

Lines AB, CD and EF have been drawn.
Which lines goes through the origin (0,0)?

Line AB.

Figure 7.5 Cartesian coordinates

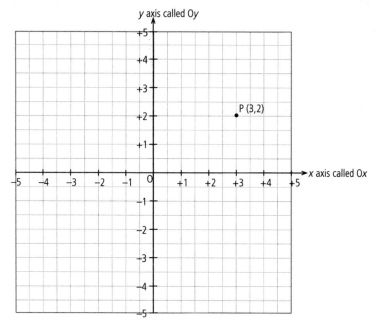

Figure 7.6 Plotting points

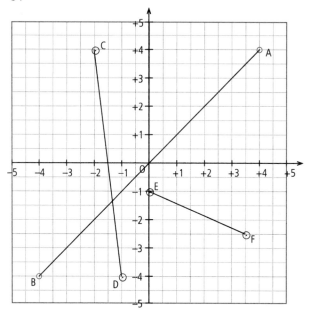

EXERCISE 7.2 Plotting points

Draw a set of axes with x and y each from −6 to +6, with a scale of 1 cm equal to 1 unit on both axes. Plot and label the following points:

A (2,6)   B (3,−5)   C (4,2)   D (−3,3)   E (0,5)   F (−1,0)   G (−2,−2)

7.3 The straight line through the origin

In Fig. 7.7 lines L, M and N all pass through the origin (0,0). They differ in their slopes. We call the slope of the line the **gradient**. It measures the **rate of change of y with x**. If y increases when x increases (lines L and M), the gradient is positive. If y decreases when x increases (line N), the gradient is negative. The gradient of line L is greater than that of line M.

Figure 7.7 Straight lines through the origin

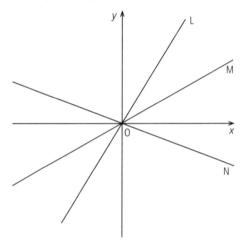

On **each** line, the ratio

$$\frac{y \text{ value}}{x \text{ value}}$$

is the same for all points. So, how can we write an equation relating y to x on a particular line through the origin? We say

$$y = mx$$

where m is always the same ('constant') for that line. m gives the value of the gradient.

Graph 1 in Fig. 7.8 shows the graph of $y = 1.5x$. On this straight line,

    coordinates of P are (2,3)
    coordinates of Q are (1,1.5)
    coordinates of R are (−2,−3)

In each case

$$\frac{y \text{ value}}{x \text{ value}} = 1.5$$

Thus, the gradient of the line is 1.5 and at every point

$$y = 1.5x$$

We call this the **equation** of the line.

Figure 7.8 Graphs of $y = 1.5x$ and $y = -1.5x$

1.
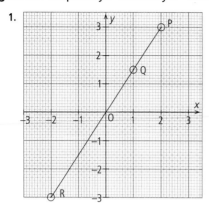

2.
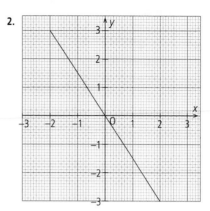

Graph 2 in Fig. 7.8 shows the graph of $y = -1.5x$. The gradient of the line is -1.5 and at every point

$$y = -1.5x$$

Summary

A straight line through the origin has the equation '$y = mx$'.

7.4   The straight line more generally

In Fig. 7.9, lines L, M and N all have the same gradient. They differ in where they cut the axes. We call the y value where a line cuts the y axis the *y* **intercept**.

The equation of line M is $y = mx$. What are the equations of lines like L and N?

Figure 7.9 Straight lines with different y intercepts

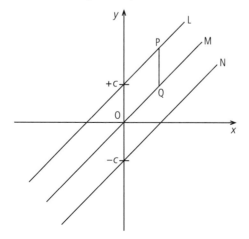

Look at the points P and Q. They have the same x value but the y value is greater at P. How much greater?

y_p at P $= y_Q$ at Q plus c

At each point on line L

y value $=$ gradient $m \times x$ value $+ c$

Thus the equation $y = mx + c$ describes line L.

The equation of any straight line not through the origin is

$y = mx + c$

It cuts the y axis at $y = c$, where x is zero. In the case of line N, the intercept is negative.

Summary

$y = mx$ represents a straight line of gradient m through (0,0).
$y = mx + c$ represents a straight line of gradient m and intercept c on the y axis.

EXAMPLE

Look at Fig. 7.10. On line M, what is y when $x = 1$?

$y = 2$ when $x = 1$

What is the equation of line M?

$y = 2x$

Where does line L cut the y axis?

At $y = 2$

What is the equation of line L?

$y = 2x + 2$

Where does line N cut the y axis?

At $y = -2$

Figure 7.10 Lines with positive gradient

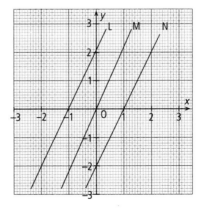

What is the equation of line N?

$y = 2x - 2$

Straight lines with negative gradient

Figure 7.11 Lines with negative gradient

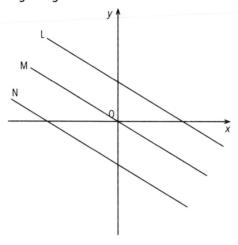

On lines L, M and N in Fig. 7.11, y decreases as x increases. The gradients are negative and all the same. The three lines, however, have different y intercepts.

EXERCISE 7.3 Equations of straight lines

Look at Fig. 7.12. **As an example we shall find the equation of line N.**

Figure 7.12 Equations of straight lines

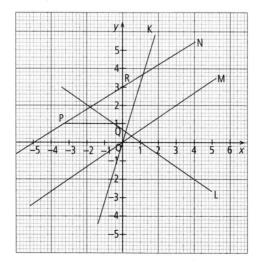

What is the gradient m? Look at triangle PQR.

$$\text{Gradient} = \frac{\text{increase in } y}{\text{increase in } x}$$

$$= \frac{RQ}{PQ}$$

$$\Rightarrow m = \frac{2}{3}$$

What is the y intercept, c?

The line cuts the y axis at $y = 3$. $\qquad\qquad c = 3$

What is the equation? [REMEMBER: '$y = mx + c$'] $y = \frac{2}{3}x + 3$

Now you can answer the following questions:
1. Which line has negative gradient?
2. What is the y intercept for line M?
3. Which lines pass through the origin?
4. Which lines have the same gradient?
5. What is the gradient of line M?
6. What is the gradient of line K?
7. What is the equation of line K?

Summary

To find the equation of a straight line:
- draw a simple triangle to find the gradient m
- read off the y intercept c
- substitute in $y = mx + c$.

7.5 How do we plot a graph when we know the equation?

If the graph is a straight line, we use **only two points**.

1. Decide which two x values to use.
2. Use the equation to calculate the corresponding y values. Now you have the coordinates of two points.
3. Select your scale.
4. Plot the two points and complete the graph.

EXAMPLES

1. Drawing $y = 2x$ for x values from -2 to $+2$.
   The line must go through the origin so one point is (0,0). When $x = 1$, $y = 2 \times 1 = 2$, so the lines goes through (1,2). On Fig. 7.13 these points have been plotted and the straight line has been drawn through them.

2. Drawing $y = x + 1$ for x values from -3 to $+3$
   Comparing with '$y = mx + c$', the y intercept must be 1. When $x = 3$, $y = 3 + 1 = 4$, so the line

Figure 7.13 Drawing a straight line using two known points

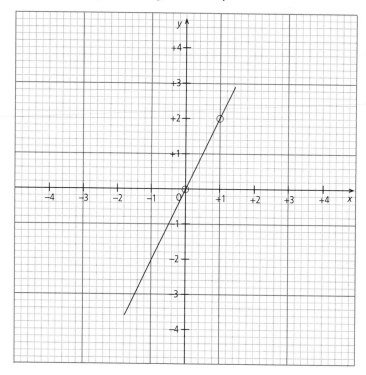

Figure 7.14 Drawing a straight line using its equation

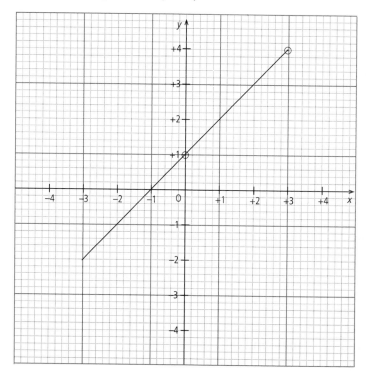

goes through (3,4). On Fig. 7.14 the points (0,1) and (3,4) have been plotted and the straight line has been drawn through them.

7.6 How do we find the equation linking the variables?

Again, we need to find the gradient 'm' and the y intercept 'c'.

The gradient is related to the size of the angle θ in in Fig., 7.15. We measure the gradient by

$$m = \frac{\text{increase in } y}{\text{increase in } x}$$
$$= \frac{y_Q - y_P}{x_Q - x_P}$$
$$= \frac{a}{b}$$

[OR the gradient $= \tan \theta$]

Figure 7.15 Finding the gradients

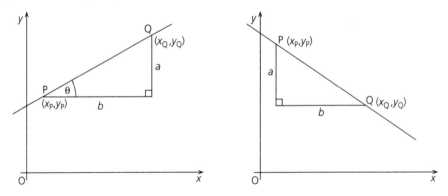

When the gradient is negative, there is a **decrease** in y for an increase in x. Then the gradient is

$$m = -\frac{a}{b}$$

The intercept c can be found by reading or calculating the y value where the line cuts the y axis. Figure 7.16 shows examples.

For the line L, $c = 0$
For the line M, $c = -1$
For the line N, $c = +2$

In experimental work, however, the numbers are not likely to be so easy. Use the data on Fig. 7.17 to find the equation of each line.

Summary

To find m, draw a triangle and calculate $\frac{a}{b}$.

Figure 7.16 Finding the *y* intercept

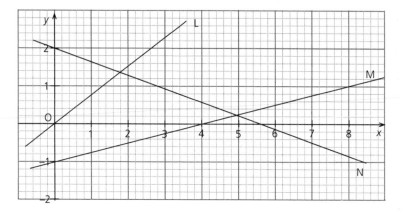

Figure 7.17 Finding the equation of a plotted line

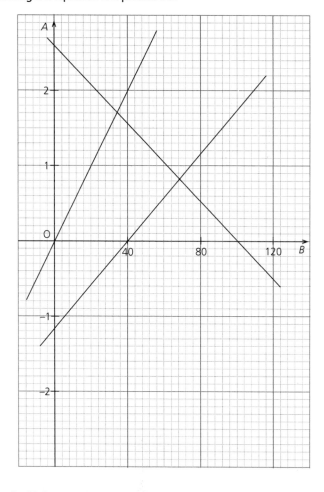

To find *c*, read off the intercept on the *y* axis.

Make sure you have the correct signs.

EXERCISE 7.4 Drawing straight lines '$y = mx + c$'

1. Draw a set of axes with x and y each from -6 to $+6$ using a scale of 1 cm equal to 1 unit on both axes.

   You are going to draw two lines on these axes.
   (a) (i) Does the line $y = 3x$ go through the origin?
       (ii) If $y = 3x$, what is y when $x = 2$?
       (iii) Use your answers to draw the line $y = 3x$ on your graph.
   (b) Another straight line is represented by $y = 2x - 1$.
       (i) What is y when $x = 0$?
       (ii) What is y when $x = 3$?
       (iii) Use your answers to draw the line $y = 2x - 1$
   (c) Read from your graph the point where the two lines $y = 3x$ and $y = 2x$ cross each other.

2. The equation of a straight line is $y = 2x + 3$
   (a) What is the gradient of the line? What is the intercept on the y axis?
   (b) Using those two answers, **sketch** the graph, roughly to scale. On the same axes, **sketch** the graph of $y = 6 - x$.

7.7 Describing relationships

We have seen in earlier chapters that relationships between variable quantities can be expressed by

- using words
- writing mathematical equations.

The nature of a mathematical relationship can also be illustrated by drawing a graph. For example, in words, 'The extension x of a spring is directly proportional to the increase in load F' is a statement of a definite relationship which can be expressed mathematically as $F = kx$, where k is a constant.

As an equation $F = kx$ is of the form $y = mx$ and represented graphically would thus be a straight line through the origin. This is illustrated by graph 1 in Fig. 7.18, where five types of relationships between two variables quantities A and B are shown.

Each of the following relationships is an example of one of these types:

- velocity and time for a uniformly accelerating body i.e. $v = u + at$ which is of the form $y = mx + c$ (graph 2).
- velocity and time for a uniformly decelerating body, i.e. $v = u + at$ where a is negative (graph 3).
- acceleration of mass acted on by a constant force i.e. $a = \frac{F}{m}$. A graph of a against m would be a reciprocal curve (graph 4).
- acceleration of mass acted on by a constant force i.e. $a = \frac{F}{m}$ where a graph of a against $\frac{1}{m}$ would be a straight line through the origin (graph 5).

Figure 7.18 Describing relationships

1.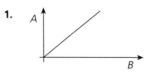

A is directly proportional to B.
Graph of the form $y = mx$.

2.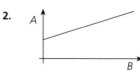

A increases linearly with B.
Graph of the form $y = mx + c$, m positive.

3.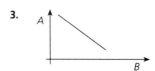

A decreases linearly with B.
Graph of the form $y = mx + c$, m negative.

4.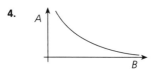

A is inversely proportional to B.
As A increases, B decreases.
Graph of the form $y = \dfrac{\text{constant}}{x}$.
Reciprocal curve.

5.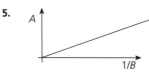

A is inversely proportional to B.
Graph of the form $y = mx$,
where $x = \dfrac{1}{B}$.

7.8 Some other important graph shapes

The parabola

This form occurs when **y depends on x²**.

Figure 7.19 The parabola

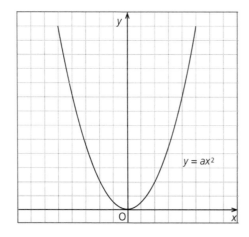

$y = ax^2$

For example, a graph of distance x against time t for a body travelling with uniform acceleration is part of a parabola.

Growth and decay curves

Figure 7.20 shows two examples.

Figure 7.20 Growth and decay curves

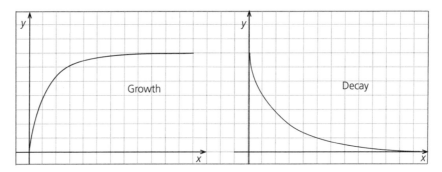

- Growth – a graph of voltage against time across a capacitor which is being charged.
- Decay – a graph of count rate against time near a decaying radioactive source.

The sine curve

As angle θ varies from 0 to 360°, the sine of the angle varies as shown in Fig. 7.21.

Figure 7.21 The sine curve

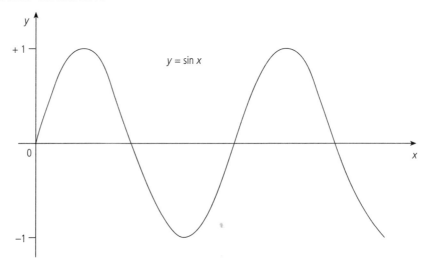

The characteristic shape is one you will have seen on the cathode ray oscilloscope screen when using it to investigate alternating voltage. Any variation following the pattern of Fig. 7.21 is said to be **sinusoidal**.

For example, the variation of current I with time t in an alternating current of mains frequency 50 Hz is given by

$$I = I_m \sin 100 \pi t$$

where I = instantaneous current at time t
    I_m = peak current

7.9 More about gradients

Earlier we used the equation $y = mx + c$ for a straight line of y plotted against x. The **gradient** m of the line tells us the **rate of change of y with x**.

That is equally true for a curve: the gradient tells us the rate of change but for a curve it is **not constant**. To find the gradient of a curve at a particular point we draw the tangent to the curve **at that point**.

Look at Fig. 7.22.

On graph 1, gradient at Q > gradient at P.
On graph 2, gradient at P is negative.
On graph 3, gradient is positive from A to B but decreasing
        gradient is zero at B
        gradient is negative from B to C and increasing
        gradient is negative from C to D and decreasing
        gradient is zero at D
        gradient is positive from D to E and increasing.

Figure 7.22 Gradients of curves

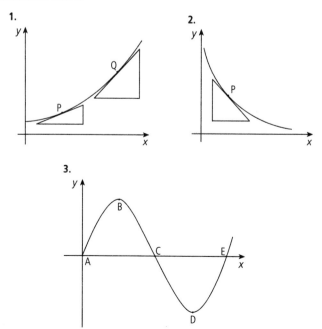

Such ideas are important in dealing with physical quantities when we need to know the rate of change.

Rate of change from gradient

A graph of quantity A against quantity B shows how A changes with B; the gradient of the graph shows the **rate of change**. For a straight line, the gradient is the same everywhere. To find its value, we measure $\tan \theta$ in Fig. 7.23.

Figure 7.23 Gradient of a line

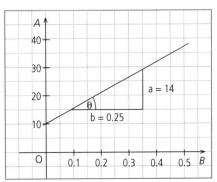

Gradient $= \tan \theta$

Gradient $= \dfrac{\text{change in } A}{\text{change in } B}$

$ = \dfrac{a}{b}$

a and b must be measured in the scales on the axes. For example in Fig. 7.23

gradient $= \dfrac{14}{0.25}$

$ = 56$

If the graph is a curve, we define its gradient as the slope of the tangent. The gradient is different at different points. For example, on the cooling curve in

Figure 7.24 Gradient of a curve is not constant

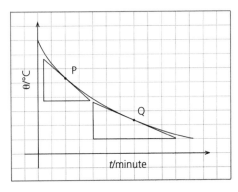

Fig. 7.24 the rate of fall of temperature decreases as the temperature falls. Note that the gradient is negative.

To measure a gradient, a tangent is drawn at the relevant point and tan θ is measured as before. The tangent must be carefully drawn and the triangle must be large enough for accurate measurement.

In Fig. 7.25. the gradient at P = tan θ

$$= \frac{a}{b}$$

$$= \frac{0.80}{1.5}$$

$$= 0.53 \text{ (2 s.f.)}$$

The rate of change of quantity A with quantity B may be another useful quantity.

Figure 7.25 Gradient of a curve at a point

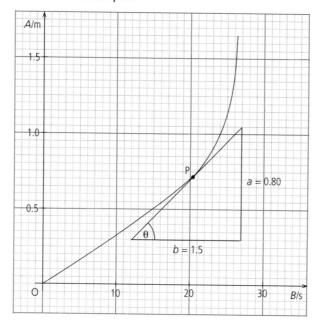

In fact, the graph in Fig. 7.25 shows the motion of a ball bearing, rolling down an inclined plane. The distance in metres has been plotted against time in seconds. From the gradient, we can find the speed at point P.

gradient = 0.53 (2 s.f.)
⇒ speed = 0.53 m/s (2 s.f.)

[NOTE: The gradient itself has no units but the speed is measured in units consistent with those on the axes.]

The shape of the graph, that is its increasing gradient, shows that the ball is accelerating.

Examples of useful quantities found from gradients are given below.

Graph	Gradient gives:
Distance against time	Speed
Temperature against time	Rate of rise or fall of temperature
Charge against time	Current
Velocity against time	Acceleration
Activity against time	Decay rate

Summary

Gradient of graph gives rate of change.

Gradient of a curve varies.

Gradient of a curve at P = gradient at tangent at P.

When measuring on a graph:
- use large triangles
- draw tangents carefully.

EXERCISE 7.5 Gradients

1. (a) What is the rate of cooling at P in Fig. 7.26?
   (b) What happens to the gradient of the curve as the time increases from 0 to 12 minutes?
   (c) Can you suggest a physical reason for that?

2. The graph in Fig. 7.27 is a distance against time graph for a moving body.
   (a) What is the gradient at P?
   (b) What physical quantity does the gradient represent?

Figure 7.26 Cooling curve: temperature θ against time *t*

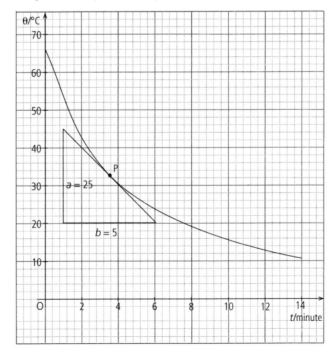

Figure 7.27 Distance against time graph

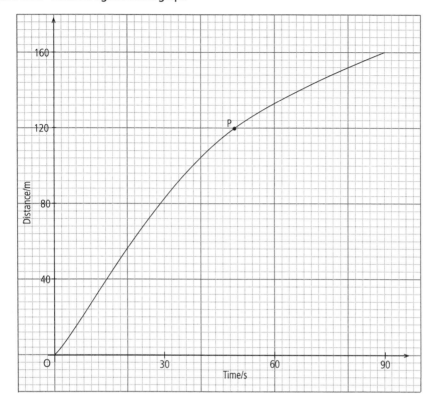

(c) What happens to the gradient as the time increases from 0 to 30 s?

(d) What does your answer to (c) mean in terms of the motion of the body?

3. The graphs in Fig. 7.28 show variations in physical quantities. In **each** case, name the quantity which is measured by the gradient of the graph.

7.10 Area between a curve and the axis

The area between the line in Fig. 7.29 and the x axis is the sum of all the areas of all the little strips like the shaded one below P on the diagram. The area of the strip PP′ is approximately $y_p \Delta x$, where Δx is the width of the strip.

To find the area between PQ on the graph and the x axis, we can add up the areas of all the strips which make up the required area, whatever the shape of the graph.

How do we calculate the area?

Whatever the method, it is essential to use the scales on the axes of the graph. We can:

Figure 7.28 Variations in physical properties

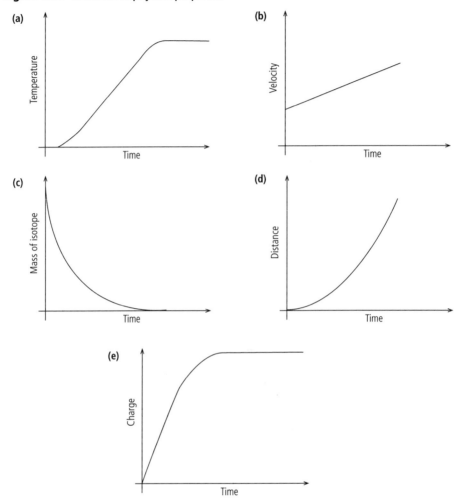

Figure 7.29 Area between a curve and the axis

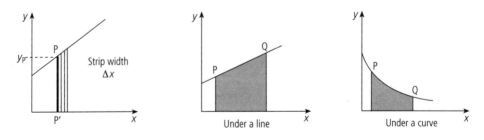

1. calculate using geometry if the shape is simple, for example a triangle or trapezium
2. count squares on the graph paper
3. calculate using an approximate rule, for example the trapezoidal rule
4. use calculus and integrate if we know the equation, but that method is beyond the scope of this book.

Why is the area useful?

There are several applications where the area on a graph relating physical quantities gives the value of another quantity. Here are some examples.

Force–extension for a spring

Looking at Fig. 7.30, which is a plot of force against extension,

Figure 7.30 Force–extension for a spring

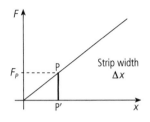 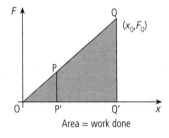

Area = work done

the area of strip PP′ $= F_p \times \Delta x$

where Δx is the change in length across the strip. This is approximately equal to the **work done**, i.e. force × distance, in stretching the spring by Δx. Adding areas of all such strips, the area shaded is the **total work done** stretching the wire. The area of the triangle OQ′Q gives

$$\text{work done} = \frac{1}{2} \text{base} \times \text{height}$$

$$= \frac{1}{2} F_Q x_Q$$

Even if the force against extension graph is not a straight line, the area under the graph still represents the work done.

Speed against time graphs

The area under a speed against time graph gives the **distance** covered, whatever the shape of the graph. Figure 7.31 shows various speed against time relationships.

Figure 7.31 Speed–time graphs

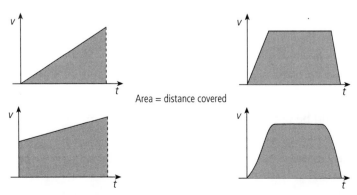

Area = distance covered

Current against time graphs

If a current I flows for a time interval Δt, the charge passed is $I\Delta t$. For longer time intervals, the charge passed can be found from the area between the graph of current against time and the t axis.

Figure 7.32 illustrates some cases. The shaded area represents the charge passed.

Figure 7.32 Graphs of current I against time t

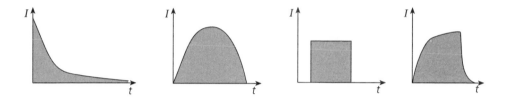

Summary

The area under a graph can represent a physical quantity.

Force against extension: area gives work done.

Speed against time: area gives distance.

Current against time: area gives charge.

7.11 Accuracy of graphs

Figure 7.33 shows the same graph plotted in different ways, assuming the points are located correctly.

Graph 1 displays the points well but at least one more point is needed to be sure of the line of best fit.

Graph 2 has scales for both A and B starting at the origin with the result that the known values are confined to a small area of the paper. It is difficult to plot them accurately on these scales and impossible to draw the best line. Perhaps it is a curve!

Graph 3 has a sensibly chosen scale for B but the A scale starts at the origin. Again the display is unsatisfactory.

Graph 4 has suitable scales and enough points, but the line is NOT the line of best fit.

Only graph 1 could be used accurately and then only if more points were added.

A graph not only displays results but is used also for reading off values. Accuracy is essential at that stage, too.

Figure 7.33 Unsatisfactory straight line graphs

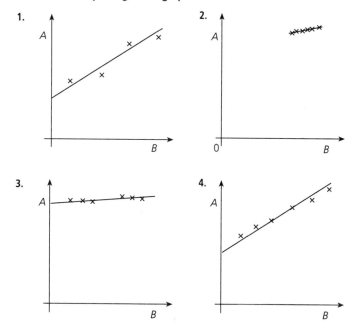

Summary

A useful graph must be accurately plotted on suitable scales.

Problems with curves

It is difficult to draw accurate curves. It is also difficult to deduce the shape and so to find the mathematical relationships between *A* and *B*. Finally, it is difficult to read off values, either within the measured range (**interpolation**) or outside that range (**extrapolation**).

Whenever possible, a relationship should be reduced to straight line form.

Information from straight lines

To find a gradient accurately, a large triangle must be drawn so that *a* and *b* are measured as precisely as possible. An 'easy' value for *b* can be chosen (e.g. 1.00 or 0.500 or 250 perhaps) but the correct number of significant figures must be stated nevertheless.

Figure 7.34 Finding the gradient

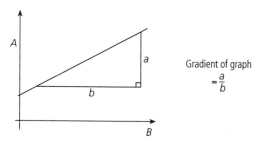

Gradient of graph

$$= \frac{a}{b}$$

It is more accurate to find the gradient of a straight line than to take the mean of a set of readings. Graphical treatment has the added advantages of exposing readings which need checking and showing deviation from linearity, say, at high values. Such anomalies otherwise would pass unnoticed.

Summary

Use straight line graphs where possible.

Graphs expose errors and unusual behaviour.

Figure 7.35 Errors and unexpected behaviour exposed by graphs

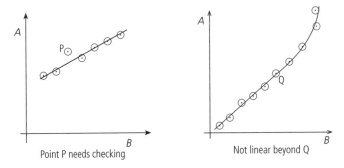

Point P needs checking   Not linear beyond Q

7.12 Nonlinear relationships

Many relationships do not follow a linear pattern. Some we can reduce to linear form but in other cases we need to gain information from a graph which is a curve.

Investigation of the relationship between current and voltage for a filament lamp does not give a straight line because the temperature of the filament rises as the current increases. Figure 7.36 illustrates the result.

Figure 7.36 Graph of voltage V against current I for a filament lamp

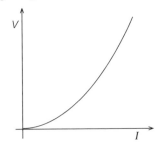

It may be useful to find the rate of change of a quantity A with a quantity B in such a case. The gradient of a curve varies, so we have to find its values at specific points. In Fig. 7.37 the rate of change of A with B at P is the gradient of the tangent at P.

In some cases, the curve has a negative gradient. For example, the graph in

Figure 7.37 Gradient of a curve

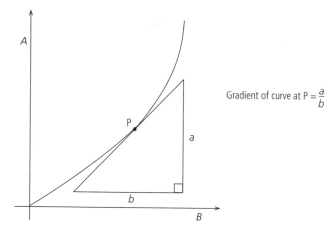

Gradient of curve at P = $\frac{a}{b}$

Fig. 7.38 shows the discharge of a capacitor. The rate of change of charge with time at P is negative.

Figure 7.38 Gradient of a curve

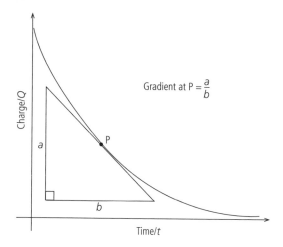

Gradient at P = $\frac{a}{b}$

Summary

If a relationship between A and B is nonlinear we can:

• reduce it to linear form
• find the rate of change of A with B from a graph.

Changing relationships to linear form

The linear equation '$y = mx + c$' includes x raised to the power one. Quadratic equations of the general form $y = ax^2 + bx + c$ contain x raised to the power two. Other relationships may involve x^3 or $\sqrt{x}$, for example.

It is often useful to change an equation to the linear form, especially when planning experimental work or plotting a graph.

EXAMPLES

1. $2ax = v^2 - u^2$, where a is constant, u^2 is constant. Rearrange in the linear form $y = mx + c$, making v^2 the subject of the formula.

   $2ax + u^2 = v^2$ (adding u^2 to both sides)
   $\Rightarrow v^2 = 2ax + u^2$

   Comparing with '$y = mx + c$'
   $\qquad\quad v^2 = 2ax + u^2$

   A graph of v^2 against x would be a straight line of gradient $2a$ and cut the v^2 axis at the value u^2.

2. $R = \dfrac{a}{V} + b$ where a and b are constants

   Comparing with '$y = mx + c$'

   A graph of R against $\dfrac{1}{V}$ would be a straight line of gradient a, cutting the R axis at the value b.

3. $T = 2\pi\sqrt{\dfrac{M + m}{k}}$ where m and k are constants

   Make T^2 the subject of the formula and rearrange in the form '$y = mx + c$'

   Squaring both sides $T^2 = 4\pi^2\left(\dfrac{M + m}{k}\right)$

   $\qquad\qquad\quad = \dfrac{4\pi^2 M}{k} + \dfrac{4\pi^2 m}{k}$

   $\Rightarrow T^2 = \dfrac{4\pi^2 M}{k} + \dfrac{4\pi^2 m}{k}$,

   comparing with '$y = mx + c$'

   A graph of T^2 against M would be a straight line of gradient $\dfrac{4\pi^2}{k}$ cutting the T^2 axis at $\dfrac{4\pi^2 m}{k}$.

EXERCISE 7.6 Relationships in linear form

Each of the following relationships can be expressed in the linear form '$y = mx$' or '$y = mx + c$'. Make the variable in [ ] the subject of the formula.

Compare with the general linear equation and state what graph you would plot to obtain a straight line.

Also state what the gradient and the 'y' intercept represent.

1. $\dfrac{1}{2}mv^2 = hf - \phi$ where h, m and ϕ are constant. $[v^2]$

2. $pV = k$ where k is constant. $[p]$

3. $v^2 = 2gh$ where g is constant. $[v^2]$

4. $\alpha = \dfrac{R - R_0}{R_0\theta}$ $[R]$

5. $x = av^2 + bv$ $[\dfrac{x}{v}]$

chapter

Using graphs

8.1 **Representing physical quantities**

When making physical measurements, we need to use appropriate units. A symbol in a formula or physical equation stands for a quantity, not a number, and so both the magnitude and the unit are part of it. Similarly, both the magnitude and the unit are involved in the scale of a graph.

On the graph of extension x against load F illustrated in Fig. 8.1, the axes are labelled F/N and x/mm which means that the figures on the axes are numbers, not quantities.

Figure 8.1 Graph of extension x against load F for a spring

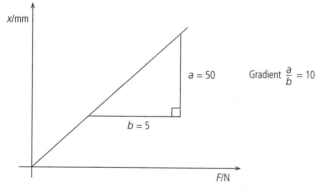

It is conventional to plot the quantity we set, in this case the load, along the horizontal axis, and the quantity we measure, in this case the extension, along the vertical axis. We use known loads and measure the extensions. The graph is a straight line as shown for a spring obeying Hooke's law. The gradient, itself a number, depends however on the units on the axes.

$$\text{Gradient} = \frac{a}{b}$$
$$= \frac{50}{5}$$
$$= 10$$

[It is essential to use the scales on the axes when finding a and b.]

The straight line graph through (0,0), of the form '$y = mx$', shows that

extension $x \propto$ load F (Hooke's law)
OR $x =$ a constant $\times F$

The constant is the **extension per unit load** and can be found from the gradient:

extension per unit load $= 10\,\text{mm/N}$

Often, the relationship between F and x is written as

$F = kx$

where k is a constant called the 'stiffness'. The stiffness is the load per unit extension and so in our experiment is measured in N/mm.

$$\text{stiffness } k = \frac{1}{\text{extension per newton}}$$
$$= \frac{1}{10}\,\text{N/mm}$$
$$= 0.1\,\text{N/mm}$$

A less stiff spring would be easier to stretch and would give a smaller answer.

We have looked at one practical example, seeing how quantities and their units are involved and how the graph can provide information.

Summary

An axis is labelled with $\dfrac{\text{quantity}}{\text{unit}}$.

Numbers are marked on the axes.

The gradient is a number.

From the gradient, a physical quantity may be found. Its unit must be stated.

EXERCISE 8.1 Graphs of linear relationships

1. The energy H supplied in heating a material through a temperature rise θ is

   $H = Mc\theta$

   where M is the mass of material and c is its specific heat capacity.
   (a) Describe the graph of H against θ.
   (b) What can be calculated from the gradient?

2. A spring is suspended and loaded over the range where it obeys Hooke's law. Its length varies according to the equation

   $l = 4M + 10$

   where l is stretched length of the spring (in cm) and M is the mass of the suspended load (in kg).
   (a) Calculate l when $M = 1$ kg and l when $M = 0$ (the unstretched length).

(b) Use your answers to plot those two points on a graph of *l* against *M*, using a range of *l* up to 15 cm. Draw a straight line through the points. [Does *l* need to start at 0?]

(c) **Use the graph** to find:
  (i) the mass which stretches the spring to 12 cm
  (ii) the length of the spring when it is loaded with 600 g
  (iii) the extension caused by a load of 800 g.

3. A rocket's speed is recorded every second after take-off.

Speed/(m/s)	8	16	24	32	40
Time/s	1	2	3	4	5

(a) Plot a speed against time graph.
(b) Describe the rocket's motion.
(c) Use the graph to find the rocket's acceleration.

4. A cyclist travels first along a flat road and later reaches a hill. Distance and time measurements for the journey are recorded.

Distance from start/(m/s)	0	40	80	180	280	380	480
Time/s	0	20	40	60	80	100	120

(a) Plot a distance against time graph.
(b) Describe the cyclist's motion on the journey.
(c) Use the graph to calculate her speed on the flat.
(d) Calculate using the graph her speed on the hill.
(e) Over what part of the journey is distance from start directly proportional to time?

8.2 Graphs used for conversions

You may need to use a **conversion graph** to change units. For example, litres can be converted to gallons and vice versa using Fig. 8.2.

From the graph:

  8 gallons = 36.4 litres   (Point P)
  10 litres = 2.2 gallons   (Point Q)

Alternatively, you may need to construct a conversion graph which converts units or which enables costs to be estimated, for example. To construct a graph based on a linear relationship, you draw a straight line through two known points, one of which may be the origin. For example, to construct the conversion graph for litres and gallons, you would know that

  10 gallons = 45.5 litres

and that the line must pass through (0,0).

Figure 8.3 illustrates another example where one point and the gradient are known. The hiring cost for some equipment is £25 plus £3 per day. The graph shows the cost for up to 14 days. It was constructed knowing that the gradient is 3 and the line must pass through (0,25).

Figure 8.2 Conversion graph: litres to gallons. This graph was constructed using the points (0,0) and (45.5,10)

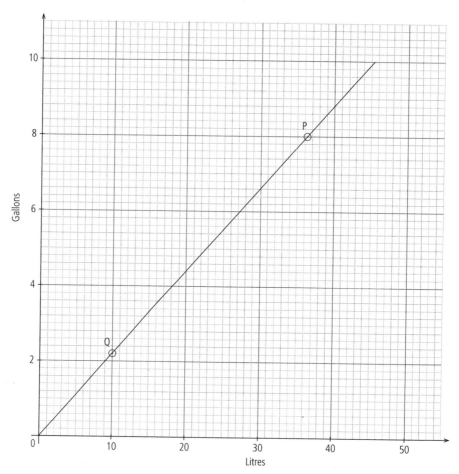

Summary

To construct a linear conversion graph:

    draw a straight line through two known points

OR   draw a line of known gradient through one known point

EXERCISE 8.2 Conversion graphs

1. An alloy is used to make cylinders. Its density is 8.89 g/cm³. Use this information to construct a graph of mass against volume for masses up to 1.0 kg.

    [HINT: Calculate volume $= \dfrac{\text{mass}}{\text{density}}$ for 1000 g.]

    Use the graph to read off:
    (a) the mass of alloy used to make a cylinder of volume 35 cm³.
    (b) the volume of a finished cylinder which has mass 0.80 kg.

Figure 8.3 Graph relating hire cost to period of hire. This graph was constructed using the point (0,25) and gradient 3

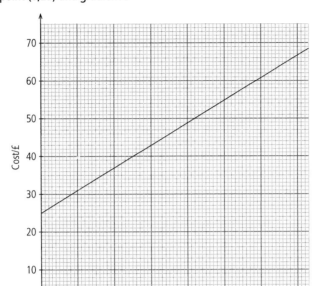

EXERCISE 8.3 Miscellaneous experimental graphs

1. In an experiment with a trolley, the braking distance of the vehicle is found for various initial speeds. After the trolley in Fig. 8.4 has covered a certain distance, the weight leaves the floor and acts as a brake.

Figure 8.4 Braking distance and speed

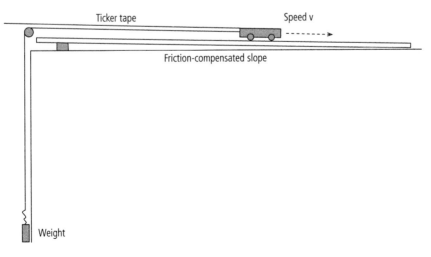

The ticker tape, which shows dots printed every $\frac{1}{50}$ s, looks like Fig. 8.5.

(a) How long a time interval is represented by 5 dot spaces? Call this time t.

Figure 8.5 Ticker tape

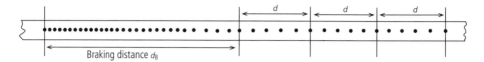

Braking distance d_B

(b) Find the average distance d covered in 5 dot spaces: use a ruler to take measurements.

(c) Calculate the initial speed $v = \dfrac{d}{t}$.

(d) Measure d_B.

(e) Results collected from other group members are given in Table 8.1. Add the result you have just calculated. Plot a graph to show how d_B varies with v.

Table 8.1 Braking distances

Speed v/(cm/s)	15.8	8.2	11.8	20.4	27.4
Braking distance d_B/cm	2.8	0.7	1.5	4.6	8.3

(f) Comment on the shape of your graph.

2. A capacitor is charged from a d.c. source through a resistor. Charge builds up on the capacitor at a rate which depends on the values of R and C and on how much charge it has gained already. The graph of charge against time is illustrated in Fig. 8.6.

Figure 8.6 Charging a capacitor

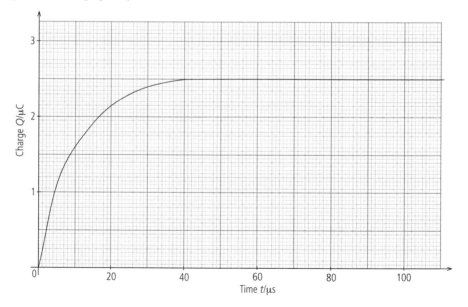

The 'time constant' τ is defined as the time for the charge to reach 0.632 of its final value Q_m.

(a) Read off the value of Q_m from the graph.

(b) Calculate $Q = 0.632\,Q_m$. Use this value of Q to read off the time constant from the graph.

(c) Use a ruler to find where the tangent to the curve at $t=0$ meets the line $Q=Q_m$. Theoretically the lines should meet where $t=\tau$. Find τ by this method.

(d) τ can be calculated without using the graph. Use the expression $\tau = RC$ to find τ if $R=2.0\,kW$ and $C=5.0\,nF$.

3. The velocity v of a body was measured at various times t as it accelerated. The results were:

t/s	1.0	4.0	7.0	10.0	13.0
$v/(m/s)$	2.6	4.4	6.2	8.0	9.8

(a) Plot a graph of v against t.
(b) Use your graph to describe the motion.
(c) Find the velocity when timing started (which is at $t=0$).
(d) Calculate the gradient of the graph.
(e) Write an equation which links v and t.

EXERCISE 8.4 Gradients of experimental graphs

[REMEMBER: Draw large triangles. Use the scales on the axes.]

1. The resistance of a thermistor varies with temperature. Use the data in Table 8.2 to plot a graph. Describe the relationship and find the rate of change of resistance with temperature at 50°C.

Table 8.2

Temperature/°C	5	10	20	30	50	80	100	120
Resistance/Ω	400	220	160	120	75	50	45	40

2. The graph in Fig. 8.7 shows the variation of rate of decay R (in grams per hour) with time (in hours) for a radioactive source.

(a) Read off the time after which the rate of decay has decreased to one half of its initial value.

(b) Find the rate of decay after 24 hours.

3. The current I flowing from a capacitor which is discharging is measured at various time intervals. The results are tabulated.

Current I/mA	220	160	70	30	10	4
Time t/ms	30	40	80	140	200	240

(a) Plot a graph of I against t.
(b) From your graph find the rate of change of current with time after 0.1 seconds from the start of the discharge.
[Think about the sign of your answer.]
(c) Write your answer in words.
['The current is decreasing ... etc.]

4. A set of results gives the readings of voltage and current for a filament lamp.

Figure 8.7 Decay rate R aginst time t for a radioactive source

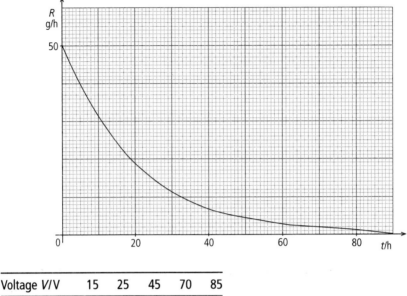

Voltage V/V	15	25	45	70	85
Current I/mA	530	592	643	674	683

(a) Plot a graph of V against I.
(b) Ohm's law says $V \propto I$. The filament lamp does not obey Ohm's law. How does V vary with I for the lamp? Why is Ohm's law not applicable?
(c) Find the rate of change of V with I when the voltage is 60 V.

EXERCISE 8.5 Graphs on the same axes

1. A computer measures the speed of a trolley travelling down an inclined plane and plots a graph of speed against time. The graph is illustrated in Fig. 8.8 for three different angles of slope of the plane.

Figure 8.8 Acceleration down a slope

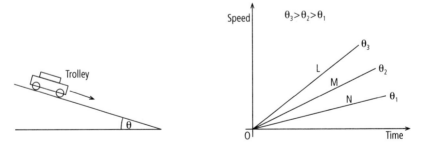

(a) What can be calculated from the gradient of line N?
(b) What happens to the gradient as angle θ is increased and why?
(c) What approximately is the greatest possible value of the gradient of lines like L, M and N?
(d) To what angle θ would that greatest possible value correspond?

8.3 Controlling variables

Think about the equation $F = ma$. It involves three quantities – force F, mass m and acceleration a – all of which can vary. To investigate the relationship methodically, we cannot vary all three quantities at the same time. What we should do is keep one variable constant. For example:

keep m constant   vary F measure a conclude $a \propto F$

OR   keep F constant   vary m measure a conclude $a \propto \dfrac{1}{m}$

From both results, we conclude that

$$a \propto \frac{F}{m}$$

Then, if 1 N is the force which gives a mass of 1 kg an acceleration of 1 m/s^2,

$$F = ma.$$

Keeping factors constant in turn is called **controlling the variables**.

In more complicated equations, some factors may be constants. For example, for a loaded cantilever, shown in Fig. 8.9:

Figure 8.9 Loaded cantilever

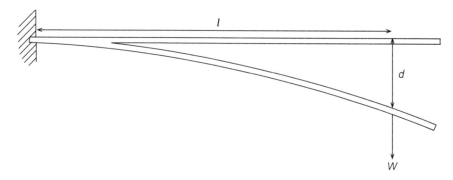

$$d = \frac{Wl^3}{3YAk^2}$$

where d is the depression where the load W is attached,
l is the distance from the clamp to the load,
W is the weight of the load,
Y is the Young modulus of the material of the beam,
A is the cross-sectional area, and
k^2 depends on the shape of the cross-section.

For a given beam, only the first three are variables. Y, A and k^2 are constants.

To find how d varies with W, we should hold l constant. Then

$$d = \text{a constant} \times W$$

and a graph of d against W should be a straight line through the origin.

Which variable must be controlled to find how d varies with l?

We must control W. Then

$$d = \text{a constant} \times l^3$$

and a graph of d against l^3 should be a straight line through the origin.

Summary

In an investigation:

- identify the variables
- hold all but two constant
- control the variables in turn.

EXERCISE 8.6 Working with a number of variables

1. The resistance R of a number of wires of different lengths l but of the same material and cross-section a is given by

$$R = \frac{\rho l}{a}$$

where ρ is the resistivity of the material.
(a) Which factors are constant?
(b) What would be the form of a graph of R against l?
(c) What could you find from the gradient of the graph?
(d) If you measured the diameter d, how could you use your answer to (c) to find the resistivity ρ?

2. A beam of breadth b and depth d is supported as shown in Fig. 8.10.

Figure 8.10 Loaded beam of rectangular cross-section

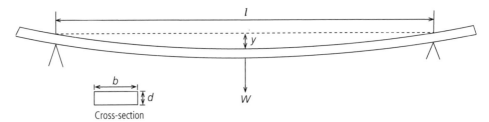

Cross-section

The depression y at the centre is measured keeping the distance l between the supports constant. The expression linking y with l and w is

$$y = \frac{kwl^3}{bd^3}$$

where k is a constant.
(a) What other factors are constant?
(b) What graph would show the relationship between y and w?
(c) What would you do to check the relationship between y and l?
   What graph would you draw?
   What result would you expect?

8.4   Families of curves

In engineering, you will often meet a set of graphs plotted on the same axes. For example, when a gas expands, pressure p, volume V and temperature T are all involved. Graphs of pressure against volume (isothermals) are drawn at different temperatures T (in kelvins). Figure 8.11 shows such a family of curves. The graphs of p against V plotted together give information about the third variable T.

Figure 8.11 Pressure against volume graphs for a gas at three different temperatures

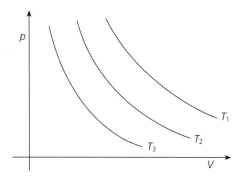

Summary

Graphs plotted on the same axes can show:

- the effect of changing a constant
- the influence of another variable.

8.5   Relationships reduced to linear form

In practice there are many relationships between physical quantities which are not linear. If we change the relationship into the form of a linear equation, we can plot a straight line graph.

For example, for uniform acceleration a from rest

$$x = \frac{1}{2}at^2$$

where x is distance and t is time.

Comparing with '$y = mx$', we could obtain a straight line by plotting x against t^2. The graph should be a straight line through the origin, with gradient $^a/_2$.

The advantages of a straight line over the curve obtained by plotting variables directly are:

1. errors and unusual behaviour are detected
2. the relationship in the form '$y = mx + c$' can be found
3. values of 'm' and 'c' are obtained.

Figure 8.12 Distance x against time t

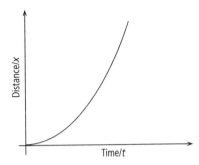

Now return to the case of graphs for uniform acceleration from rest. Distance x against time t gives a curve as shown in Fig. 8.12, while x against t^2 gives a straight line through (0,0). The acceleration a can be found from the gradient of the graph in Fig. 8.13.

Figure 8.13 Graph of x against t^2

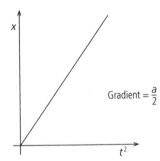

Examples of nonlinear equations which can be converted to linear form are listed in Table 8.3.

EXERCISE 8.7 Reducing to linear form

1. The stopping distance x_s of a car after the driver notices an obstacle is given by

$$x_s = av^2 + bv$$

where v is the speed.
(a) Rewrite the equation in linear form.
(b) Values of x_s for a particular driver and vehicle are:

x_s/yd	5	14	26	42	61	84
v/m.p.h.	10	20	30	50	50	60

Plot a graph to enable you to find a and b. a will be in yd/(m.p.h.)^2 and b in yd/m.p.h.
(c) Change the units so that a is in kg/N by multiplying your value by 4.57. Then change b into seconds by multiplying your value by 2.04.
(d) Which constant is related to the reaction time of the driver and which to the braking force of the car?
(e) Estimate what the stopping distance might be at 70 m.p.h.
(f) Comment on your answers to (d) and (e).

Table 8.3 Reduction to linear form

Relationship	Linear form $'y = mx + c'$	Graph to plot	Through (0,0)?	Gradient $'m'$	Intercept $'c'$
$x = \frac{1}{2}at^2$	$x = \frac{a}{2}t^2$	x against t^2	yes	$\frac{a}{2}$	0
$2ax = v^2 - u^2$	$v^2 = 2ax + u^2$	v^2 against x	no	$2a$	u^2 on v^2 axis
$v^2 = 2gh$	$v = \sqrt{2gh}$ or $v^2 = 2gh$	v against $\sqrt{h}$ v^2 against h	yes yes	$\sqrt{2g}$ $2g$	0 0
$\Delta p = 2kv^2/\rho$	$\Delta p = \text{constant} \times v^2$	Δp against v^2	yes	$2k/\rho$	0
$R = \frac{a}{V} + b$	$R = a\left(\frac{1}{V}\right) + b$	R against $\frac{1}{V}$	no	a	b on $\frac{1}{V}$ axis
$pV = k$	$p = k\frac{1}{V}$	p against $\frac{1}{V}$	yes	k	0
$\frac{1}{2}mv^2 = hf - \phi$	$v^2 = \frac{2h}{m}f - \frac{2\phi}{m}$	v^2 against f	no	$\frac{2h}{m}$	$-\frac{2\phi}{m}$
$T = 2\pi\sqrt{\frac{M+m}{k}}$	$T = \frac{4\pi^2 M}{k} + \frac{4\pi^2 M}{k}$	T^2 against M	no	$\frac{4\pi^2}{k}$	$\frac{4\pi^2}{k}$
$x_s = av + bv^2$	$\frac{x_s}{v} = bv + a$	$\frac{x_s}{v}$ against v	no	b	a

<h2 style="background:black;color:white;display:inline">8.6</h2> **Calculating a physical quantity from the area under a graph**

You saw in Chapter 7 that the area under a graph can represent a physical quantity. In this chapter we shall make calculations using the three methods outlined in the Summary.

Summary

To find area in terms of the scales on a graph:

1. calculate using geometry
2. count squares
3. use the trapezoidal rule.

We shall apply those methods to practical examples.

Calculation using geometry

When the graph comprises straight lines, calculation is easy.

Figure 8.14 Work done in stretching a wire

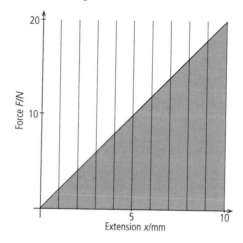

In Fig. 8.14 a wire is stretched by 10 mm.

Work done = area of shaded triangle

$$= \frac{1}{2} \text{force} \times \text{extension}$$

$$= \frac{1}{2} \times 20 \times \frac{10}{1000} \text{J}$$

$$= 0.10 \text{ J}$$

Figure 8.15 Distance covered from a graph of velocity against time

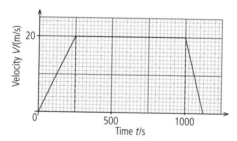

Figure 8.15 is a graph of velocity against time for a journey.

$$\text{Total distance covered} = \left(\frac{1}{2} \times 250 \times 20\right) + \left(750 \times 20\right) + \left(\frac{1}{2} \times 125 \times 20\right) \text{m}$$

$$= 18\,750 \text{ m}$$

Area by counting squares

Figure 8.16 is a graph of current against time. The charge passed in the positive half cycle of an alternating current can be estimated. First, in terms of the scales on the axes, the area of a square of convenient size is found. Then the number of squares is counted, starting with whole squares and then adding in the part squares. Use a tally if it helps.

The area in Fig 8.16 can be split into two equal halves. Then, counting the squares in each half of the diagram:

Figure 8.16 Area by counting squares: graph of current I against time t

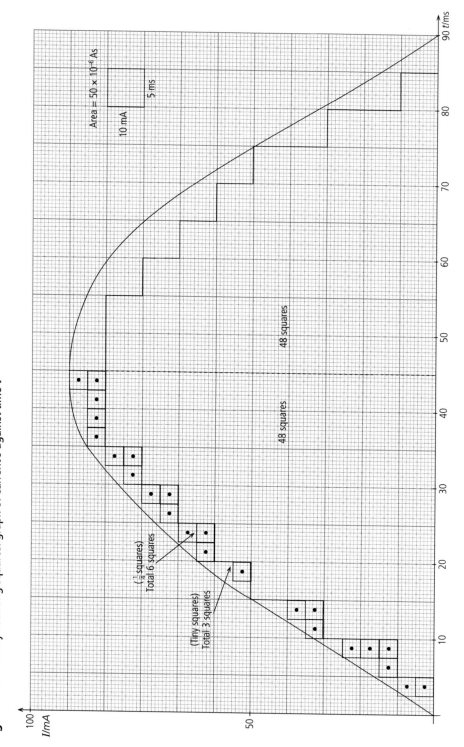

1. Count whole squares. Total = 48 cm².
2. Count quarter squares. Total 24 = 6 cm².
3. Estimate remaining tiny squares. Total ≈ 3 cm².

Final number of squares for whole diagram = 2(48 + 6 + 3) = 114.

[There are many ways of doing this. Choose your own method.]

The number assessed on the graph is 114, so the charge Q passed is given by

$$Q = \text{number} \times \text{area of a square.}$$
$$= 114 \times 50 \times 10^{-6} \text{ As}$$
$$= 5.7 \times 10^{-3} \text{ coulombs}$$

If we used calculus to integrate

$$I = I_m \sin 2\pi f t$$

where f is the frequency, we should obtain

$$Q = 5.73 \times 10^{-3} \text{ coulombs (3 s.f.)}$$

which is the correct answer. Counting squares can only give an approximate answer.

EXERCISE 8.8 Area under a graph

1. The graph in Fig. 8.17 illustrates a train journey between two stations.

Figure 8.17 A train journey: graph of velocity against time

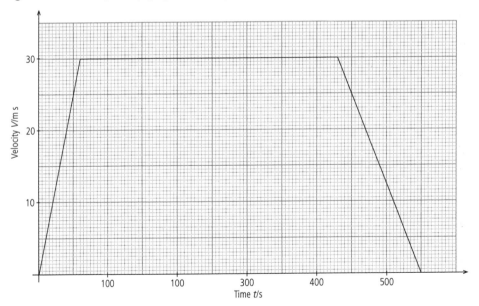

(a) What is the train's acceleration at the beginning?
(b) What is the train's deceleration at the end?
(c) What is the train's highest speed?
(d) How long does the train take to make the journey?
(e) How far is it between the two stations?

2. A rubber band is stretched by adding weights one by one up to an extension of 150 mm. Then the weights are removed one by one. The graphs for loading and unloading follow different curves.

Figure 8.18 Stretching a rubber band

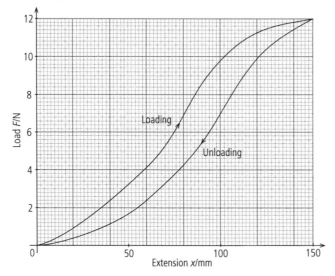

(a) By counting squares on the graph in Fig. 8.18, estimate the total work done in loading the rubber.

(b) Is the potential energy recovered when the band is unloaded more or less than your answer?

(c) Can you suggest what happens to the missing energy?

The trapezoidal rule

The area to be calculated is divided into a number, n, of strips of equal width Δx. Each strip approximates to a trapezium of area

$$\left(\frac{y_a + y_b}{2}\right) \Delta x$$

as illustrated in Fig. 8.19.

Figure 8.19 The trapezoidal rule

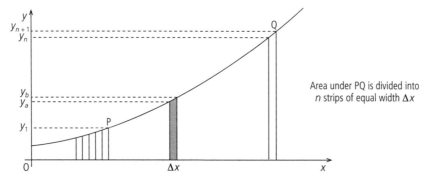

Area under PQ is divided into n strips of equal width Δx

The total area between P and Q is therefore the sum of the areas of the n strips.

$$\text{Area} = \frac{\Delta x}{2}\left[(y_1+y_2)+(y_2+y_3)+\ldots+(y_n+y_{n+1})\right]$$

$$= \Delta x\left[\frac{(y_1+y_{n+1})}{2}+y_2+y_3+\ldots+y_n\right]$$

The greater the number n of strips used, the more accurate the answer.

In Fig. 8.20 the half-period sine curve is divided in 2×18 strips. Here

$$y_1 = 0 \quad y_{n+1} = 100\,\text{mA} \quad \frac{y_1+y_{n+1}}{2} = 50.0\,\text{mA}$$

$$\sum_{y_2}^{y_n} y = 8+16+25+34+41+50+58+65+71+77+82+87+91+94$$
$$+97+98+99 = 1093\,\text{mA}$$

The trapezoidal rule gives

$$\text{Area} = 2\times\Delta x\left[\frac{y_1+y_{n+1}}{2}+\sum_{y_2}^{y_n}y\right]$$

$$= 2\times 2.5\,[50+1093]\times 10^{-6}\,\text{C}$$
$$= 5.72\times 10^{-3}\,\text{C}$$

Again the value is approximate, not exact.

Summary

The trapezoidal rule for finding area under a graph is

Area = width of strip × [average of first and last y values + sum of all other y values]

8.7 Finding the mean value

The area under a graph of a varying quantity helps us to find its average (mean) value.

For example, the accurate value for the total charge passed in the positive half cycle of alternating current in Fig. 8.20 is $5.73 \times 10^{-3}\,\text{C}$. To calculate the mean current, we use

$$\text{mean current} = \frac{\text{total charge}}{\text{total time}}$$

$$= \frac{5.73\times 10^{-3}}{90\times 10^{-3}}\,\text{A}$$

$$= 63.7\,\text{mA}$$

The mean current is related to the peak current I_m by

$$I_{\text{average}} = 0.637\,I_m$$

The average over a whole cycle would be zero because the current reverses.

Figure 8.20 Area by trapezoidal rule: graph of current I against time t

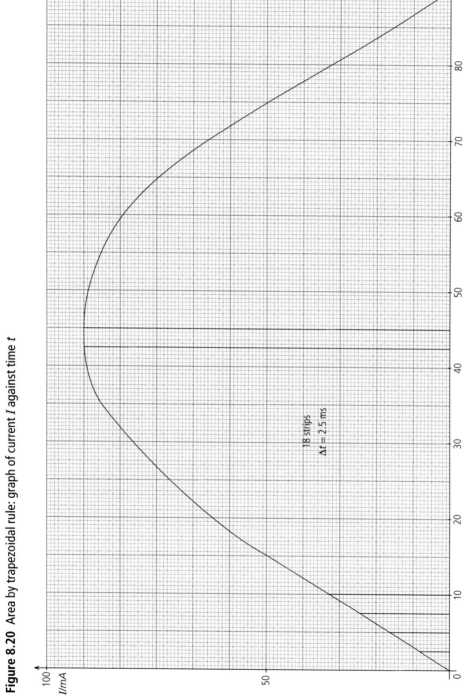

18 strips
$\Delta t = 2.5$ ms

Figure 8.21 Mean square current: graph of I^2 against t where peak value $I_m = 1$ A

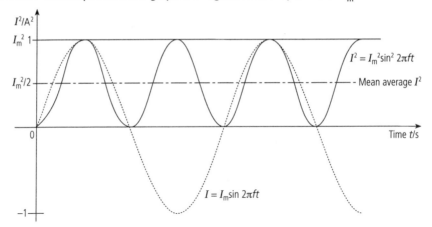

Neither the peak current nor the half cycle average is the most important value. What matters most is the **effective value**, which determines the heating effect. To calculate it, we need the mean square current. The graph of I^2 against t is shown in Fig. 8.21. The average value of I^2 is halfway up the curve; that is $\dfrac{I_m^2}{2}$.

The **root mean square (r.m.s.) current** is $\dfrac{I_m}{\sqrt{2}}$.

Summary

For sinusoidal alternating current

$$I = I_m \sin 2\pi f t$$

where I_m is the peak value and f is the frequency.

Effective value $I_{r.m.s.} = \dfrac{I_m}{\sqrt{2}} = 0.707\,I_m$

Half cycle average value $I_{av} = \dfrac{2}{\pi} I_m = 0.637\,I_m$

In general, the mean value of a quantity varying with x can be found from

$$\text{mean value} = \frac{\text{total area under graph}}{\text{range of } x}$$

as illustrated in Fig. 8.22.

Figure 8.22 Mean value

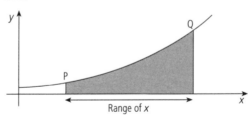

Average y over the range from P to Q $= \dfrac{\text{shaded area}}{\text{range of } x}$

EXERCISE 8.9 Applications of area under a graph

1. When a capacitor discharges through a resistor, the current variation with time is as shown
   in Fig. 8.23.

Figure 8.23 Graph of current I against time t during discharge of a capacitor

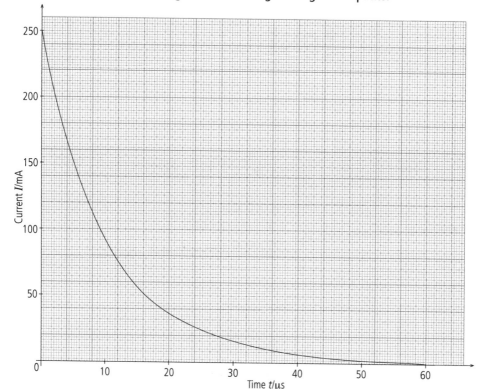

Use the graph to find approximately the original charge on the capacitor, using the
trapezoidal rule.

[HINT: Original charge = total charge passed through the resistor]

2. An alternating current varies with time in a square waveform of amplitude 0.4 mA and
   frequency 200 Hz.
   (a) Sketch the graph of current against time.
   (b) What is the half cycle average value of current? [QUICKLY!]
   (c) What charge passes in a half cycle?
   (d) What is the r.m.s. value of the current? [QUICKLY!]

Self-assessment test

1. $8 + 13 + 5$ is
   - (a) 24
   - (b) 25
   - (c) 26
   - (d) 27

2. Add
   3
   11
   7   is
   - (a) 19
   - (b) 20
   - (c) 21
   - (d) 22

3. $7 - 4 - 1$ is

      (a) 2

      (b) 4

      (c) –2

      (d) 12

4. $-7 + 8$ is

      (a) –1

      (b) 1

      (c) 15

      (d) –15

5. 62.00 ± 0.03 lies between

      (a) 59.00 and 65.00

      (b) 61.70 and 62.30

      (c) 61.97 and 62.03

      (d) None of these

6. 210×3.5 is

      (a) 213.5

      (b) 630.5

      (c) 735

      (d) 740

7. $8 \div 0.16$ is

      (a) 0.5

      (b) 2

      (c) 20

      (d) 50

8. 0.3×0.3 is

      (a) 0.09

      (b) 0.9

      (c) 0.6

      (d) None of these

9. $\dfrac{7}{16}$ correct to 2 decimal places is

      (a) 0.43

      (b) 0.44

      (c) 0.49

      (d) None of these

10. $\dfrac{9}{16} + \dfrac{5}{64}$ is

      (a) $\dfrac{13}{16}$

      (b) $\dfrac{23}{32}$

      (c) $\dfrac{43}{64}$

      (d) $\dfrac{41}{64}$

11. The square of 16 is
> (a) 32
> (b) 64
> (c) 4
> (d) 256

12. 9^2 is
> (a) 81
> (b) 18
> (c) $4\frac{1}{2}$
> (d) 3

13. 25 is the square of
> (a) 625
> (b) 50
> (c) $12\frac{1}{2}$
> (d) 5

14. $\sqrt{1600}$ is
> (a) 40
> (b) 80
> (c) 160
> (d) 400

15. The square root of 600 is between
> (a) 10 and 20
> (b) 20 and 30
> (c) 30 and 40
> (d) 40 and 50

16. The square root of 0.9 is
> (a) 0.3
> (b) 0.81
> (c) Between 0.9 and 1.0
> (d) None of these

17. The side of a square of area 400 mm² is
> (a) 20 mm
> (b) 40 mm
> (c) 63 mm
> (d) 200 mm

18. A rectangle 80 mm by 20 mm has the same area as a square of side
> (a) 40 mm
> (b) 50 mm
> (c) 400 mm
> (d) 1600 mm

19. If the sides of two squares are in the ratio 2:3 the ratio of their areas is
> (a) 2:3
> (b) 4:9
> (c) 4:25
> (d) 9:25

20. If the diameters of two circles are in the ratio 1:2 the ratio of their areas is
    - (a) $1:\pi$
    - (b) 1:2
    - (c) 1:4
    - (d) 2:3

21. In $1\,m^2$ there are
    - (a) $10\,mm^2$
    - (b) $100\,mm^2$
    - (c) $1000\,mm^2$
    - (d) $1\,000\,000\,mm^2$

22. In $1\,m^2$ there are
    - (a) $10^6\,mm^2$
    - (b) $10^3\,mm^2$
    - (c) $10^2\,mm^2$
    - (d) $10^1\,mm^2$

23. 0.000011×100 is
    - (a) $0.000\,000\,11$
    - (b) 0.0011
    - (c) 0.011
    - (d) 100.000\,011

24. 0.04×10^3 is
    - (a) 0.12
    - (b) 1.2
    - (c) 4
    - (d) 40

25. The mean of 15, 17, 18.25, 15.75 is
    - (a) 15.75
    - (b) 16
    - (c) 16.5
    - (d) 66

26. 5°36′ written in decimals is
    - (a) 5.36°
    - (b) 5.60°
    - (c) 0.536°
    - (d) None of these

27. In Fig. 1, if the angle at C is 35°15′, then the angle at A is
    - (a) 35°15′
    - (b) 25°
    - (c) 24°45′
    - (d) None of these

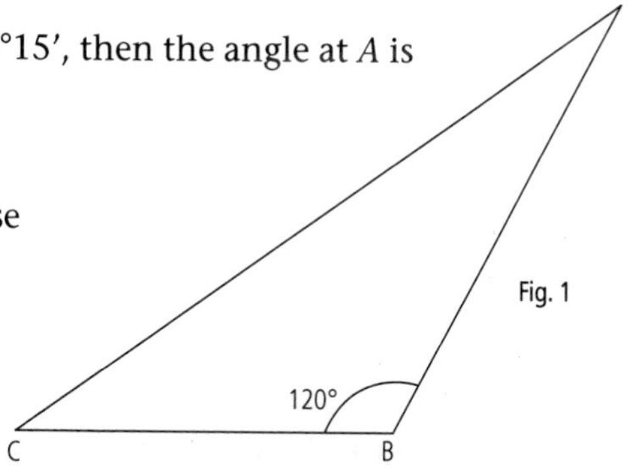

Fig. 1

28.

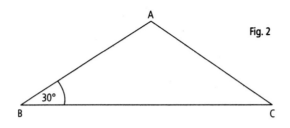

Fig. 2

In Fig. 2, if side *AB* is equal to side *AC*, then the angle at *A* is
     (a) 30°
     (b) 60°
     (c) 90°
     (d) 120°

29.

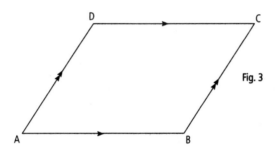

Fig. 3

In Fig. 3, side *AB* is parallel to side *DC*, and side *AD* is parallel to side *BC*.
If the angle at *A* is 60°, then the angle at *D* is
     (a) 30°
     (b) 60°
     (c) 120°
     (d) None of these

30.

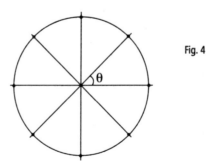

Fig. 4

In Fig. 4, if the points marked on the circle are equally spaced, then angle
θ is
     (a) 30°
     (b) 45°
     (c) 60°
     (d) None of these

31. If $20 + 120 + x = 180$, then x is
    (a) 10
    (b) 40
    (c) 140
    (d) 320

32. If $0.16x = 8$, then x is
    (a) $\frac{1}{2}$
    (b) 2
    (c) 50
    (d) None of these

33.

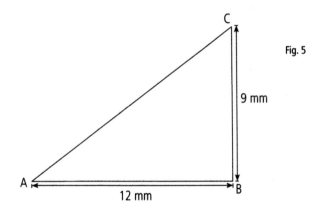

Fig. 5

In Fig. 5, if the angle at B is a right angle, then side AC is
    (a) 12 mm
    (b) 15 mm
    (c) 21 mm
    (d) None of these

34.

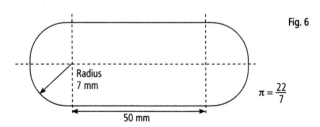

Fig. 6

In Fig. 6, the perimeter of the figure is
    (a) 100 mm
    (b) 122 mm
    (c) 144 mm
    (d) 254 mm

35.

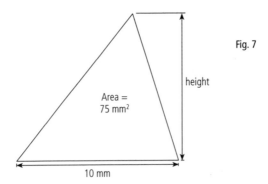

Fig. 7

height

Area =
75 mm²

10 mm

Given area = $\dfrac{\text{base} \times \text{height}}{2}$, the height of the triangle in Fig. 7 is

    (a) 7.5 mm
    (b) 15 mm
    (c) 37.5 mm
    (d) None of these

36. The diameter of the base of a cone is 30 mm. The area of the base is
    (a) $15\pi^2 \, \text{mm}^2$
    (b) $30\pi \, \text{mm}^2$
    (c) $225\pi \, \text{mm}^2$
    (d) $900\pi \, \text{mm}^2$

37. Two cylinders of the same material and having equal cross-sectional areas have heights in the ratio $2:3$. The ratio of their masses is
    (a) $2:3$
    (b) $4:9$
    (c) $8:27$
    (d) None of these

38. A brass contains copper and zinc in the proportion of 7 parts copper and 3 parts zinc by mass. If the brass contains 39 g of zinc the total mass of the brass is
    (a) 117 g
    (b) 91 g
    (c) 130 g
    (d) None of these

39. The percentage of copper in the brass in question 38 is
    (a) 7%
    (b) 50%
    (c) 70%
    (d) None of these

40.

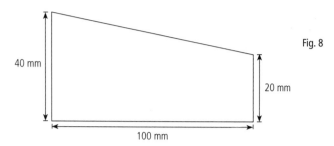

Fig. 8

40 mm

20 mm

100 mm

The surface area of the sheet metal template in Fig. 8 is
- (a) 2000 mm²
- (b) 3000 mm²
- (c) 4000 mm²
- (d) None of these

41. Given $\pi = \frac{22}{7}$, the length of sheet metal required to form a cylinder of diameter 210 mm (ignore thickness) is
- (a) 660 mm
- (b) 650 mm
- (c) 640 mm
- (d) None of these

42. If $R = 4$, $I = 3$ then I^2R is
- (a) 24
- (b) 36
- (c) 48
- (d) 144

43. If $\frac{1}{R} = \frac{1}{2} + \frac{1}{6}$ then R is
- (a) $\frac{1}{8}$
- (b) 8
- (c) $\frac{2}{3}$
- (d) $1\frac{1}{2}$

44. If $R = \frac{V}{I}$ then V is
- (a) $\frac{R}{I}$
- (b) $\frac{I}{R}$
- (c) IR
- (d) None of these

45. If $P = \dfrac{V^2}{R}$ then V is

    (a) $\dfrac{PR}{2}$

    (b) $\dfrac{P-R}{2}$

    (c) $2PR$

    (d) $\sqrt{PR}$

46. Given that resistance $= \dfrac{\text{Voltage}}{\text{Current}}$,

an appliance carrying a current of 0.16 amperes with a voltage of 8 volts has a resistance of

    (a) 1.28 ohms

    (b) $\dfrac{1}{2}$ ohms

    (c) 50 ohms

    (d) None of these

47.

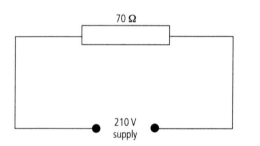

Fig. 9

Given that resistance $= \dfrac{\text{Voltage}}{\text{Current}}$,

in Fig. 9 the current through the resistor is

    (a) 3 amperes

    (b) $\dfrac{1}{3}$ amperes

    (c) 140 amperes

    (d) None of these

48. For an increase in temperature of 1°C a 1 mm length of steel will expand 0.000011 mm. For an increase in temperature of 5°C, a 20 mm length of steel will expand by

    (a) 0.000 055 mm

    (b) 0.000 22 mm

    (c) 0.0011 mm

    (d) None of these

49.

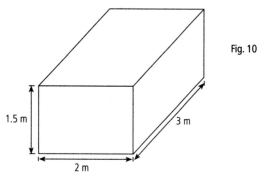

Fig. 10

1.5 m

3 m

2 m

The volume of the solid shown in Fig. 10 is
  (a) 6.5 m³
  (b) 9 m³
  (c) 13 m³
  (d) 19.5 m³

50. A twist drill running at 600 rev/min has a feed rate of 0.15 mm/rev. The feed per second is
  (a) 0.015 mm
  (b) 1.5 mm
  (c) 15 mm
  (d) 90 mm

Answers

Chapter 2 Answers

Exercise 2.1

(a) 1 is 1 hundred thousand i.e. 100 000
    0 is no ten thousands   i.e. 0
    1 is 1 thousand   i.e.   1 000
    2 is 2 hundreds   i.e.   200
    6 is 6 tens   i.e.   60
    7 is 7 units   i.e.   7
                                 101 267

(b) 3 is 3 ten thousands   i.e.   30 000
    9 is 9 thousands   i.e.   9 000
    0 is no hundreds   i.e.   0
    5 is 5 tens   i.e.   50
    0 is no units   i.e.   0
                                 39 050

Exercise 2.2

Table 2.3 Decimal fractions in words and numbers

Words	T	U	.	t	h
1.	1	7	.	2	
2.	7	1	.	1	4
3.		3	.	2	0
4.		5	.	0	7
5.		0	.	0	1
6. Eleven and twenty hundredths					
7. Four and seven hundredths					
8. Ten and three hundredths					
9. Sixty-nine hundredths					
10. Five hundredths					

Exercise 2.3 Place value

1. $40+7+\dfrac{6}{10}+\dfrac{3}{100}$

2. $100+0+8+\dfrac{6}{10}+\dfrac{9}{100}$

3. $0+\dfrac{1}{10}+\dfrac{0}{100}+\dfrac{2}{1000}$

4. $1000+0+10+2+\dfrac{6}{10}+\dfrac{9}{100}$

5. $600+10+8+\dfrac{2}{10}+\dfrac{0}{100}$

Exercise 2.4 Relative size of decimal fractions

1. 0.8, 0.81, 0.88, 0.89, 0.9
2. 0.0450, 0.43, 0.495, 0.5, 1.05

Exercise 2.5 Multiplication of decimal fractions

(a) 0.72   (b) 0.49
(c) 0.25   (d) 0.06
(e) 0.02   (f) 0.006
(g) 0.0012   (h) 0.018
(i) 5.0   (j) 0.9

Exercise 2.6 Division of decimal fractions

(a) 2   (b) 2
(c) 2   (d) 0.5
(e) 0.5   (f) 0.3
(g) 0.1   (h) 2
(i) 50   (j) 500

Exercise 2.7 Relative sizes of fractions

1. $\dfrac{1}{9},\ \dfrac{1}{7},\ \dfrac{1}{6},\ \dfrac{1}{4},\ \dfrac{1}{3}$
2. 0.11, 0.14, 0.17, 0.25, 0.33

Exercise 2.8 Equal common fractions and lowest terms

1. $\dfrac{2}{10},\ \dfrac{3}{15},\ \dfrac{4}{20},\ \dfrac{5}{25},\ \dfrac{6}{30}$
2. $\dfrac{1}{4},\ \dfrac{1}{8},\ \dfrac{1}{5},\ \dfrac{3}{4},\ \dfrac{1}{2}$

Exercise 2.9 Addition of common fractions

(a) $1\frac{1}{4}$, 1.25   (b) $\frac{59}{64}$, 0.92

(c) $1\frac{8}{15}$, 1.53   (d) $\frac{7}{18}$, 0.39

(e) $\frac{7}{8}$, 0.88

Exercise 2.10 Multiplying and dividing common fractions

(a) $\dfrac{3}{20}$ (b) $\dfrac{5}{9}$ (c) $\dfrac{2}{3}$

(d) $\dfrac{15}{16}$ (e) $1\dfrac{1}{4}$ (f) $\dfrac{27}{32}$

(g) $3\dfrac{3}{4}$ (h) 1 (i) $1\dfrac{9}{16}$

(j) 6 (k) $1\dfrac{1}{3}$ (l) $\dfrac{9}{40}$

Exercise 2.11 Estimation and the calculator

1. (a) Estimation $(30+30) \times 20 = 1200$
   Calculator 1080

   (b) Estimation $(20+20) \times \dfrac{800}{10} = 3200$

   Calculator 2736

   (c) Estimation $(30-10) \times 50 + (40-20) = 20 \times 50 + 20 = 1020$
   Calculator 757

   (d) Estimation   either $(170-50) \times 20 \div (60 \times 10)$
   $$= \dfrac{120 \times 20}{600}$$
   $$= 4$$
   OR $(200-50) \times 20 \div (60 \times 10)$
   $$= \dfrac{150 \times 20}{600}$$
   $$= 5 \text{ (less accurate)}$$
   Calculator 4

   (e) Estimation
   (not necessary) $\dfrac{(20 \times 10) - (10 \times 20)}{20 \times 20} = \dfrac{200 - 200}{400}$
   $$= 0$$

   NOTE: You should realize that 17×13 is the same as 13×17.

   (f) Estimation $\dfrac{(20+20) \times 100}{3} = \dfrac{40 \times 100}{3}$
   $$= \dfrac{4000}{3}$$
   $$\text{approx. } 1300$$
   Calculator 1369

   (g) Estimation $\dfrac{(170+30-20) \times 20}{4}$ [NOTE: $88 \div 22 = 4$]
   $$= \dfrac{180 \times 20}{4}$$
   $$= 900$$
   Calculator 1035

(h) Estimation $\dfrac{(250 \times 3 \times 20)}{350 \times 10} = \dfrac{30}{7}$

   Calculator   4   approx. 4

NOTE: Quick method $\dfrac{240 \times 3 \times 17^1}{20\ 340 \times 9\ _3} = \dfrac{24}{6} = 4$ (not all cancelling shown)

2. Calculator shows 0. Multiplying by zero always give the answer zero.
3. Calculator may show E (for Error). This means that you cannot divide by nothing! Mathematically we say the answer for division by zero is so big that we call it an infinite number (the symbol is ∞).

Exercise 2.12 Standard form

1. 30 000 000   300 000   300 000 000
   3×10^7 3×10^5 3×10^8
   in order: 3×10^5 3×10^7 3×10^8
2. 150 000   500 000   3 000
   1.5×10^5 5×10^5 3×10^3
   in order: 3×10^3 1.5×10^5 5×10^5
3. 9×10^9 2×10^6 3×10^4
   in order: 3×10^4 2×10^6 9×10^9
4. 3.313×10^3 km.
5. 1 300 000 000; 1.3×10^9 years ago.
6. 8 600 000; 8.6×10^6 km^2.
7. 6.33×10^{-7} m.

Exercise 2.13 Powers of ten

(a) $10 \times 10 \times 10$ $\times$ $10 \times 10 = 10^5$
(b) 10 $\times$ $10 \times 10 \times 10 = 10^4$
(c) 10×10 $\times$ $10 \times 10 = 10^4$
(d) $\dfrac{10 \times 10 \times 10}{10 \times 10} = 10^1$ i.e. 10
(e) $\dfrac{10 \times 10 \times 10 \times 10 \times 10 \times 10 \times 10}{10 \times 10 \times 10 \times 10 \times 10} = 10^2$
(f) $\dfrac{10 \times 10 \times 10 \times 10 \times 10}{10 \times 10}$ $\times$ $10 \times 10 \times 10 = 10^6$

Exercise 2.14 Powers of ten

Same answers as Exercise 2.13

Exercise 2.15 Factorizing

(a) $3^2 \times 2^4$ (b) 13^2
(c) $3^2 \times 5^2$ (d) $2^6 \times 5^3$
(e) $3^3 \times 2^3 \times 5^3$ (f) $2^9 \times 5^3$
(g) $2^3 \times 5^3$ (h) $3 \times 2^3 \times 5^3$
(i) 5^3 (j) 2^6

Exercise 2.16 Squares, cubes and roots

Answers are provided in the tables.

Table 2.4 Squares and square roots

Number n	Number n written as square	Square root $\sqrt{n}$
	4×4 or 4^2	4 or $\sqrt{16}$
	5×5 or 5^2	5 or $\sqrt{25}$
	6×6 or 6^2	6 or $\sqrt{36}$
	10×10 or 10^2	10 or $\sqrt{100}$
	12×12 or 12^2	12 or $\sqrt{144}$
	13×13 or 13^2	13 or $\sqrt{169}$
	15×15 or 15^2	15 or $\sqrt{225}$
	25×25 or 25^2	25 or $\sqrt{625}$

Table 2.5 Cubes and cube roots

Number n	Number n written as cube	Cube root $\sqrt[3]{\ }$
	$3 \times 3 \times 3$ or 3^3	3 or $\sqrt[3]{27}$
	$4 \times 4 \times 4$ or 4^3	4 or $\sqrt[3]{64}$
	$5 \times 5 \times 5$ or 5^3	5 or $\sqrt[3]{125}$
	$10 \times 10 \times 10$ or 10^3	10 or $\sqrt[3]{1000}$

Exercise 2.17 Estimating square roots

Table 2.6

Number $n < 1$	$\sqrt{n}$ exact or estimated
0.60	between 0.7 and 0.8
0.70	between 0.8 and 0.9
0.81	0.9
1.00	1.00
0.94	between 0.9 and 1.0
0.04	0.2
0.09	0.3
0.12	between 0.3 and 0.4

Exercise 2.18 Finding reciprocals

1. $\dfrac{1}{2}, \dfrac{1}{19}, \dfrac{1}{25}, \dfrac{1}{n}, \dfrac{1}{R}$

2. $2, 3, \dfrac{4}{3}, \dfrac{7}{2}, n$

3. $f, \dfrac{V}{I}, \dfrac{n}{2}, \dfrac{V}{Q}, \dfrac{1}{\omega t}, \sqrt{\dfrac{g}{l}}$

Exercise 2.19 Directed numbers

1. (a) +5   (b) −4
   (c) 0   (d) −4
   (e) +10   (f) −10
   (g) +6   (h) +6
   (i) −21   (j) −21
2. $(+20) - (+14) = +6$
3. $(-5) + (-10) + (+20) = +5°C$

Exercise 2.20 Square roots and indices

1. Table 2.7

Integer	+ve $\sqrt{}$	−ve $\sqrt{}$
+9		
+16	+4	−4
+25	+5	−5
+49	+7	−7
+64	+8	−8
+81	+9	−9
+100	+10	−10
+400	+20	−20
+625	+25	−25
+900	+30	−30

2. (a) 3^{-4} (b) 5^{10}
   (c) 2^{-5} (d) 3^2
   (e) 3   (f) 5^{-9}
   (g) 10^{-7} (h) 10^6
   (i) −8   (j) $(-3)^3$

Exercise 2.21 A mixed collection

1. $70\,000\,000$ km, 7×10^7 km
2. Estimation $\dfrac{30 \times 10 \times 10 \times 100}{10^2} = 3000$
   Calculator 2252.61
3. $t = 2$ [in mathematics it is +2 or −2, but −2 does not make sense in this context.]
4. $0 + (+3) + (+2) - (+15)$
5. 70×10^9 Pa
6. 50 ohms $(50\,\Omega)$
7. $\dfrac{3}{100}$
8. (a) 12   (b) 12

9. (a) Estimation $\dfrac{10 \times 60}{20} = 30$; student's answer incorrect.

    Calculator 25.3

10. (a) 3.1×10^5 (b) 1.85×10^6

    (c) 3.3   (d) 1.3×10^{-3}

    (e) 3.92×10^{-7}

11. 7.364×10^9 km

12. $\dfrac{1}{50}, \dfrac{3}{4}, \dfrac{1}{8}, \dfrac{3}{8}, \dfrac{27}{40}$

13. 0.2, 0.4, 0.6, 0.8, 0.04, 0.12, 0.2, 0.4, 0.3, 0.5

14. 1.5, 1.25, 2.6, 3.75, 5.04, 1.12

15. (a) 0.0625   (b) 0.125

    (c) 0.0625   (d) 0.25

    (e) 0.3

    Order 0.0625, 0.125, 0.25, 0.3

16. 20, 300, 400, 100, 250

Exercise 2.22 Using a calculator

1. Estimation $2 \times 3 \times \sqrt{\dfrac{0.5}{10}} = 6 \times \sqrt{0.05}$

                             approx. 6×0.25

                             $= 1.5$

    Calculator 1.45

2. Estimation $\sqrt{\dfrac{450}{3}} = \sqrt{150}$ approx 12

    Calculator 11.85

3. Estimation $\dfrac{400}{2 \times 3}$ approx. 67

    Calculator 65.25

4. Estimation $200 \times (20 + 20) - 200 \times 10 = 6000$

    Calculator 4872

5. Estimation $3 \times 100 \times 15 = 4500$

    Calculator 5033.36

6. Estimation $\dfrac{10 \times 6}{30 + 20} = \dfrac{60}{50}$ approx. 1

    Calculator 1.73

7. Estimation $\dfrac{1}{3} \times 49 \times 10$ approx. 160

    Calculator 136.61

8. Estimation $\dfrac{3}{6} \times 6^3 = 3 \times 36$

                      $= 108$

    Calculator 96.98

9. Estimation $5 \times 2 \times 2 = 20$

    Calculator 14.81

10. Estimation $\dfrac{600 \times 4}{3 \times 100} = 8$

    Calculator 6.90

NOTE: Your estimations may be slightly different from those above.

Chapter 3 Answers

Exercise 3.1 Ratio

1. (a) $6:7$
   (b) $7:13$
   (c) $6:13$
2. $\dfrac{1}{2}$ metre
3. $1:50\,000$
4. $1:200$
5. 2 ohms
6. (a) $9:1$
   (b) $9:10$

Exercise 3.2(a) Direct proportion

1. 210 miles
2. A 210 mL, B 180 mL
3. 24.0 N
4. A £7500, B £4500
5. 8 mm

Exercise 3.2(b) Inverse proportion

1. a is 3.75
2. $0.019\,\text{m}^3$

Exercise 3.3 Finding percentages

1. (a) £100 (b) £1500
2. (a) 70 marks (b) 35 marks
3. (a) 5 components (b) 23 components
4. 90%
5. 380 seats

Exercise 3.4 Changing fractions to percentages

1. 25%, 80%, 100%
2. 32.5%, 67.5%, 93.2%
3. 10%, 4%, 2.5%
4. 50%, 25%, 12.5%

Exercise 3.5 Changing percentages to fractions

1. (a) $0.1, \dfrac{1}{10}$ (b) $0.05, \dfrac{1}{20}$ (c) $0.08, \dfrac{2}{25}$ (d) $0.25, \dfrac{1}{4}$
2. (a) $0.075, \dfrac{3}{40}$, (b) $0.575, \dfrac{23}{40}$ (c) $0.33, \dfrac{33}{100}$

Exercise 3.6 Mixed percentages

1. £4 and £12.50
2. 825 mph
3. (a) 30 (b) 75%

4. (a) 300% (b) 200%
5. (a) £1300 (b) £227.50
6. 5.45%
7. 31.60%
8. Alloy composition: copper 1.925 kg, zinc 0.98 kg, nickel 0.595 kg
9. $\dfrac{1}{8}, \dfrac{3}{10}, \dfrac{17}{20}$
10. 0.375, 0.01, 0.005
11. 56.67%, 42%, 56%
    The first is just best
12. 7.5%, 150%, 26.$\dot{6}$%

Exercise 3.7 Significant figures and decimal places

1. 64.8 (3 s.f.)
2. 6.0 (2 s.f.)
3. 0.28 (2 s.f.)
4. 0.000 62 or 6.2×10^{-4} (2 s.f.)
5. 31.1 (1 d.p.)

Exercise 3.8 Tolerance

1. (a) $1\,\text{k}\Omega \pm 10\%$ (b) $330\,\text{k}\Omega \pm 2\%$
2. (a) Green, blue, red, red (b) Yellow, violet, yellow, silver
   (c) Blue, grey, green, gold
3. $544\,\mu\text{F}, 816\,\mu\text{F}$
4. 3%, 10.3 mm and 9.7 mm
5. (a) Lowest $x = 4.5\,\text{mm}$ (b) max $z - $ min $y = 1.6\,\text{mm}$
6. $\pm 0.5\%$

Exercise 3.9 Mixed questions

1. 0.000 055 m
2. 298 mm
3. $1:2$
4. 50%
5. (a) $2:9:9$ (b) $3.6\,\text{m}^3$ OR $3\dfrac{3}{5}\,\text{m}^3$
6. 11.97 lies between 11.98 and 11.90, therefore shaft is satisfactory

Chapter 4 Answers

Exercise 4.1 Tally chart and frequency tables

1.

Group	Frequency
1–4	9
5–9	8
10–13	5
14–17	4
18–21	4

2.

Dish	Tally	Frequency
Beefburgers	ЖТ ЖТ ЖТ ЖТ I	21
Chicken and chips	ЖТ ЖТ ЖТ ЖТ ЖТ	25
Sandwiches	ЖТ ЖТ ЖТ IIII	19
Beans on toast	ЖТ ЖТ I	11

Exercise 4.2 Interpreting a pictogram

1. (a) 2050   (b) In 1995   (c) 1990 to 1991

Exercise 4.3 Bar charts

1.

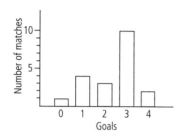

2.

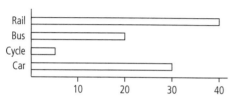

3.

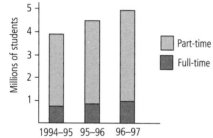

Exercise 4.4 Pie charts

1.

L & T	H & SC	E	BS
120°	105°	54°	81°

2.

Wages	Rent	Utilities	Other
180°	90°	45°	45°

3.

Black	White	Other
324°	25.2°	10.8°

4.

Australia	W. Europe	Americas	Others
144°	108°	90°	18°
£24M	£18M	£15M	£3M

Exercise 4.5 Line graphs

1.(a)

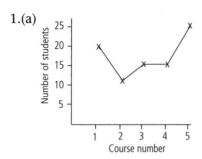

(b) A bar chart
   – visually more impact
   – not showing a trend
   – not part of a whole.

2.

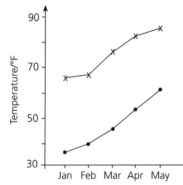

Exercise 4.6 Histograms and frequency polygons

1.

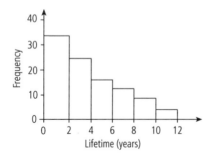

2.

Class/minutes	1.55–	1.95–	2.35–	2.75–	3.15–	3.55–	3.95–	4.35–4.75
Frequency	1	2	6	9	11	7	3	1

2. (a) and (b)

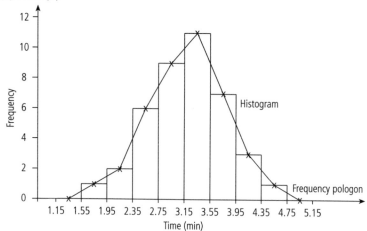

2. (c)

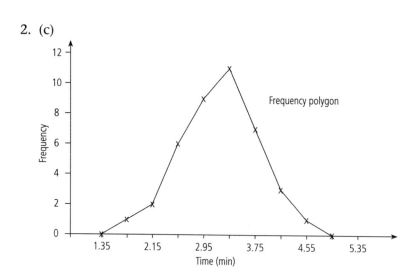

2. (d) About 90% of times are between 2.35 min and 4.35 min.
Most frequent time is about 3.35 min.
Curve leans towards lower times.

Exercise 4.7 Cumulative frequency

1. (a) 56%   (b) 9 failed   (c) 5 gained distinction
2. (a)

Resistance group/kΩ	−4.0	−4.2	−4.4	−4.6	−4.8	−5.0	−5.2	−5.4
Frequency	3	11	26	50	75	89	98	100

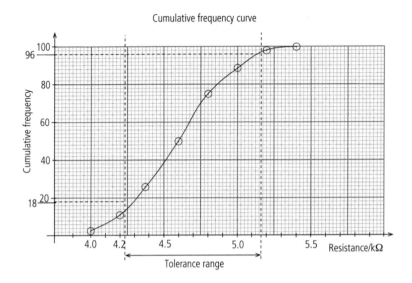

Cumulative frequency curve

(b) 5.17 kΩ, 4.23 kΩ   (c) Tolerance range marked.
(d) 13 + 4 = 17 (Percentiles at 13 and 96): 17 components per 100

Exercise 4.8 Arithmetic mean

1. 450
2. 27 (to nearest 1)
3. [When calculating, ignore the 60s: $\dfrac{1+0+3+2+5}{5} = 2.2$]

Average price = 62p to nearest 1p.
4. 4.60 kΩ (3 s.f.)

Exercise 4.9 Median and mode

1. 5
2. 59p and 61p
3. £12 500

Exercise 4.10 Simple probabilities

1. (a) $\dfrac{1}{6}$ (b) $\dfrac{1}{2}$

2. (a) $\dfrac{1}{52}$ (b) $\dfrac{1}{4}$

3. (a) $\dfrac{1}{20}$ (b) $\dfrac{19}{20}$

4. (a) $\dfrac{37}{45}$ (b) $\dfrac{8}{45}$ (c) $\dfrac{37}{44}$

Exercise 4.11 Addition and multiplication rule

1. $\dfrac{8}{27}$

2. $\dfrac{2}{13}$

3. $\dfrac{1}{240}$

4. $\dfrac{1}{12}$

Exercise 4.12 Gantt charts

1. (a)

Week no.	1	2	3	4	5	6	7	8	9
No. of men	1	1	1	6	6	5	3	2	2

(b)

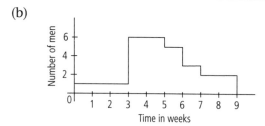

(c)

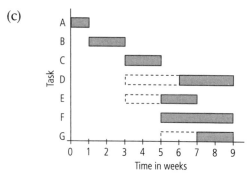

Week no.	1	2	3	4	5	6	7	8	9
No. of men	1	1	1	3	3	3	5	5	5

(d)

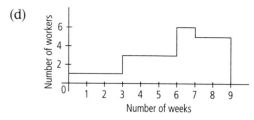

Early start gives decreasing use of manpower which could be deployed elsewhere. With late start, five are still needed. Perhaps a further modification of start times could help more.

Chapter 5 Answers

Exercise 5.1 Relationships of the form $y = mx$

Equation $y = mx$	y (variable)	x (variable)	m (constant)
Example $V = RI$	V (voltage)	I (current)	R (resistance)
(a) $s = ut$	s (distance)	t (time)	u (constant speed)
(b) $m = \rho v$	m (mass)	v (volume)	ρ (density)
(c) $Q = mL$	Q (quantity of heat)	m (mass)	L (specific latent heat)
(d) $p = \rho gh$	p (pressure)	h (height)	ρg (density × g)

Exercise 5.2 Working with like terms

Expression	Simplified
1. $2u - 4u + 3$	$-2u + 3$
2. $-6t + 2t + 3t$	$-t$
3. $7t^2 + a - 2t^2$	$5t^2 + a$
4. $5at + 7 - 4at$	$at + 7$
5. $3a - 4a + 1$	$-a + 1$
6. $3R - 4R^2 + 2$	cannot simplify
7. $6t + t^2 - 6t$	t^2
8. $\frac{3}{2}I + I - I^2$	$\frac{5}{2}I - I^2$
9. $3u \div u$	3
10. $2I^2 \div I$	$2I$
11. $15t \div 3t^2$	$\frac{5}{t}$
12. $4I \times 2I$	$8I^2$
13. $\frac{3a}{2} \times \frac{2a}{3}$	a^2
14. $4I \div 2I \times I$	$2I$
15. $6x \div x^2 \times x$	6

Exercise 5.3 Multiplying and dividing powers

Expression	Working	Simplified
2. $u^7 \div u^4$		u^3
3. $a^{-3} \times 2a^3$	$2a^0$	2
4. $(2a)^2 \times (2a)^3$	$4a^2 \times 8a^3$	$32a^5$
5. $u^{-2} \div u^{-2}$		1
6. $(3x)^2 \div \sqrt{x} \times x^{3/2}$	$9x^2 \div x^2$	9
7. $(a^2)^{1/2} \times (b^3)^{1/3}$	$a^1 \times b^1$	ab
8. $\sqrt{x} \times \sqrt{x} \times x\sqrt{x}$	$x \times x^{3/2}$	$x^{5/2}$

Exercise 5.4 Using brackets

1. $x(y+a)$
2. $I(R_1+R_2)$
3. $x(xy+1)$
4. $3a(a+2b)$
5. $R=R_0(1+\alpha\theta)$
6. $2a^2+2ab$
7. $a^2-3ab-a^2=-3ab$
8. $x^2+xy-x^2+xy=+2xy$
9. $2a-9a-b=-7a-b$
10. $x=tu+\dfrac{1}{2}at^2$

Exercise 5.5 More practice with brackets

1. $5a+5b-15$
2. $4a+a^2$
3. x^2-2x
4. $-10a+5a^2-5a^3$
5. x^2-8xy
6. $-21+26I_2$
7. $5R_1+20R_2$
8. $4t+13t^2$

Exercise 5.6 Addition and subtraction of algebraic common fractions

1. $\dfrac{7x}{12}$

2. $\dfrac{4a+b}{10}$

3. $\dfrac{6a-b}{16}$

4. y

5. $\dfrac{5}{x}$

6. $\dfrac{5y-3x}{xy}$

7. $\dfrac{17p}{16}$ OR $1\dfrac{1}{16}p$

8. $\dfrac{5}{6a}$

9. $\dfrac{13x}{21}$

10. $\dfrac{15y-4x}{6xy}$

Exercise 5.7 Multiplying and dividing algebraic common fractions

1. $\dfrac{2}{5}$

2. $\dfrac{a^2}{5}$

3. $2\dfrac{b}{a}$

4. $\dfrac{7b}{y}$

5. x^2

6. $\dfrac{4}{3}$

7. $\dfrac{16y}{a}$

8. $\dfrac{3a}{4}$

9. $(3a)^2$

10. $2a$

Exercise 5.8 Substitution

Expression	Working	Simplified
1. $2a + t$	$2 \times 10 + 3$	23
2. $-5a + 3t$	$-5 \times 10 + 3 \times 3$	-41
3. $3at^2 - 2t$	$3 \times 10 \times 9 - 2 \times 3$	264
4. $\dfrac{a^2 - 1}{t} - \dfrac{1}{t}$	$\dfrac{100}{3} - \dfrac{1}{3} = \dfrac{99}{3}$	33
5. $5t - 3a$	$5 \times 3 - 3 \times 10$	-15
6. $5t + \dfrac{1}{2}at^2$	$5 \times 3 + \dfrac{1}{2} \times 10 \times 9$	60
7. $2\pi rh$	$2\pi \times 4 \times 5$	40π
8. πr^2	$\pi \times 4 \times 4$	16π
9. $2\pi r(r + h)$	$2\pi \times 4 \, (4 + 5)$	72π
10. $\pi r^2 h$	$\pi \times 16 \times 5$	80π

Exercise 5.9 Forming equations

1. $C = h + nx$
2. $N = G - D$
3. $y = 2x$
4. $y = 2x + 3$
5. $P = 2(l + b)$
6. $A = l^2$
7. $W = Fd$
8. $V = RI$

Exercise 5.10 Solving equations

1. $x = 3$ 2. $y = -2$
3. $v = 2$ 4. $x = 13$
5. $x = 5$ 6. $x = 25$
7. $t = \dfrac{10}{3}$ OR $3\dfrac{1}{3}$ 8. $t = \dfrac{10}{3}$ OR $3\dfrac{1}{3}$
9. $t = \dfrac{10}{3}$ OR $3\dfrac{1}{3}$ 10. $t = \dfrac{40}{3}$ OR $13\dfrac{1}{3}$

Exercise 5.11 Solving equations with brackets

1. $x = 3$ 2. $x = 1$
3. $x = \dfrac{1}{2}$ 4. $x = 18$
5. $a = 7$ 6. $a = 3$
7. $R = \dfrac{4}{3}$ 8. $R = \dfrac{3}{4}$
9. $a = \dfrac{14}{5}$ OR $2\dfrac{4}{5}$ 10. $y = -2$

Exercise 5.12 Solving and checking equations

1. $x = +11, -11$ 2. $a = +4, -4$

3. $a = +2, -2$

4. $p = +\dfrac{5}{2}, -\dfrac{5}{2}$

5. $x = 7$

6. $a = 8$

7. $x = 2$

8. $x = -7$

Exercise 5.13 Changing the subject of a formula

Formula	Working (if required)	Solution
1. $V = RI$ (R)	÷ b.s. by I	$R = \dfrac{V}{I}$
2. $C = \pi d$ (π)		$\pi = \dfrac{C}{d}$
3. $W = FS$ (S)		$S = \dfrac{W}{F}$
4. $A = lb$ (l)		$l = \dfrac{A}{b}$
5. $V = kT$ (k)		$k = \dfrac{V}{T}$
6. $A = \pi dh$ (h)	÷ b.s. by πd	$h = \dfrac{A}{\pi d}$
7. $P = I^2 R$ (R)		$R = \dfrac{P}{I^2}$
8. $A = \pi r^2$ (r)	$\dfrac{A}{\pi} = r^2$	$r = \sqrt{\dfrac{A}{\pi}}$
9. $F = \sigma A$ (σ)		$\sigma = \dfrac{F}{A}$
10. $\omega T = 2\pi$ (ω)		$\omega = \dfrac{2\pi}{T}$

Exercise 5.14 More examples in transposing

1. $\theta = \dfrac{R - R_0}{R_0 \alpha}$

2. $R_0 = \dfrac{R}{1 + \alpha\theta}$

3. $I = \dfrac{V}{R}$

4. $\omega = \dfrac{2\pi}{T}$

5. $V = \dfrac{C}{Q}$

6. $g = \dfrac{4\pi^2 l}{T^2}$

7. $\theta = \dfrac{l - l_0}{l_0 \alpha}$

8. $l_0 = \dfrac{l}{1 + \alpha\theta}$

Exercise 5.15 Substituting in formulae

1. $v = 470$ 2. $E = 900$
3. $V = 36\pi$ 4. $V = 250\pi$
5. $k = 5.4 \times 10^4$

Exercise 5.16 Rearranging the subject of a formula

1. $I = \sqrt{\dfrac{E}{Rt}}$

2. $u = \sqrt{v^2 - 2as}$

3. $d = \sqrt[3]{\dfrac{6V}{\pi}}$

4. $l = \dfrac{T^2 g}{4\pi^2}$

5. $g = \dfrac{4\pi^2 l}{T^2}$

6. $v = \sqrt{\dfrac{2k}{m}}$

7. $y = \dfrac{Wl^3}{3dAk^2}$

8. $l = \sqrt[3]{\dfrac{3dyAk^2}{W}}$

9. $k = \sqrt{\dfrac{Wl^3}{3dyA}}$

10. $k = \dfrac{4\pi^2(M+m)}{T^2}$

11. $X = \sqrt{Z^2 - R^2}$ 12. $R = \sqrt{Z^2 - X^2}$

Chapter 6 Answers

Exercise 6.1 Calculation of perimeters

1. 32 cm   2. 36 m   3. 26 mm
4. 34 cm   5. 70 m   6. $2a + 2b + c$
7. 22 m   8. 44 m   9. 84 m

Exercise 6.2 Calculation of areas

1. 22 m²   2. 126 cm²   3. 175 mm²

4. 36π m²   5. 47.75 m²   6. $\left(322 - \dfrac{23}{2}\pi\right)$m²

Exercise 6.3 Calculation of surface areas

1. (a) 5400 mm²   (b) 2400 mm²
2. (a) 6000π mm²   (b) 7250π mm²
3. (a) 120 mm   (b) 800π mm²
4. 600π mm²

Exercise 6.4 Calculations of volumes

1. 300 m³
2. 480 000 mm³
3. 3850 m³
4. 90 000 mm³
5. 109 000 mm³
6. 7.06×10^{-1} kg
7. 15 000π mm³
8. 28 080π mm³

Exercise 6.5 Finding angles made by straight lines

1. $a = 45°$ (∠s on straight line)
2. $a = 40°$ (∠s on straight line)
3. $b = 80°$ (vertically opposite)
   $a = 100°$ (straight line)
   $c = a = 100°$ (vertically opposite)
4. $c = 65°$ (vertically opposite)
   $a = 115°$ (straight line)
   $b = 115°$ (straight line)
   $d = 65°$ (alternate ∠s)
   $e = a = 115°$ (corresponding ∠s)
   $f = 65°$ (corresponding ∠s)
   $g = b = 115°$ (corresponding ∠s)
5. $x = 60°$ (alternate ∠ equal to ∠ on straight line)

Exercise 6.6 Finding angles made by parallel lines

$d = 50°$ (isosceles △)
$c = 130°$ (straight line)
$e = 80°$ (∠ sum of △)
$f = 100°$ (straight line)
$a = d = 50°$ (alternate ∠s)
$b = e = 80°$ (alternate ∠s)

Exercise 6.7 Calculation of angles in a parallelogram

1. Parallelogram
   $x = 32°$ (32° alternate ∠s)
   $c = 120°$ (opposite ∠s in a parallelogram)
   $y = 180° - 120° - 32°$ (∠ sum of △)
   $\quad = 28°$
2. Rectangle
   $a = 45°$ (isosceles right-angled △)
   $b = 45°$ ($a + b = 90°$ rectangle)
   $c = 180° - 45°$ (straight line)
   $\quad = 135°$
3. Rhombus
   $a = 40°$ (isosceles △)
   $c = 50°$ (diagonals at right angles)
   $b = c = 50°$ (alternate ∠s)
NOTE: Your reasons may be different from those above.

Exercise 6.8 Finding angles in a circle

1. $a = 40°$ (on same arc)
   $b = 21°$ (on same arc)
2. $a = 38°$ (∠ sum of right-angled △)
3. $a = 180° - 65°$ (opp. ∠s of cyclic quadrilateral)
   $\quad = 115°$
   $b = 70°$ (opp. ∠s of cyclic quadrilateral)

Exercise 6.9 Using Pythagoras's theorem

1. Length of strut = 1.73 m
2. Projection = 2.18 m
3. $h = 5.75$
4. $a = 10.4$
5. $17^2 = 289$; $15^2 + 8^2 = 289$
   θ must be 90°

Exercise 6.10 Converting degrees and radians

Degrees	Radians with π	Degrees (nearest °)	Radians (2 d.p.)
60	$\dfrac{\pi}{3}$	60	1.05
		120	2.09
80	$\dfrac{4\pi}{9}$	240	4.19
		300	5.24
100	$\dfrac{5\pi}{9}$	30	0.52
		80	1.40
120	$\dfrac{2\pi}{3}$	100	1.75
		200	3.49
240	$\dfrac{4\pi}{3}$	360	6.28
300	$\dfrac{5\pi}{3}$		
360	2π		

Exercise 6.11 Finding trigonometric ratios

1.

$\theta°$	$\sin\theta$	$\cos\theta$
0	0	1.00
10	0.17	0.98
30	0.50	0.87
45	0.71	0.71
60	0.87	0.50
80	0.98	0.17
90	1.00	0

Observe that:
$\sin 0° = \cos 90°$
$\sin 90° = \cos 0°$
$\sin 45° = \cos 45°$
$\sin 30° = \cos 60°$
$\sin 60° = \cos 30°$

$\sin\theta° = \cos(90 - \theta)$
for $0 \le \theta \le 90°$

2.

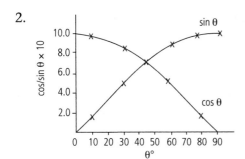

3.

θ°	tan θ
0	0
10	0.18
20	0.36
30	0.58
40	0.84
45	1.00
60	2
70	3
80	6
85	11
89	57
89.5	115
89.9	573
90	Er2

Observe that for:
- $0 \le \theta \le 45°$, tan θ increases from 0 to 1
- $\theta = 45°$, $\tan \theta = 1 \left(\dfrac{\sin 45°}{\cos 45°} = \dfrac{0.71}{0.71} = 1 \right)$
- $89 \le \theta \le 90°$, tan θ increases rapidly
- $\theta = 90°$, one calculator reads Er2, which means division by zero was attempted
  $\left(\dfrac{\sin 90°}{\cos 90°} = \dfrac{1}{0}, \therefore \tan 90° \to \infty \right)$

Exercise 6.12 Finding the angle given the trigonometric ratio

1. $\theta = 0°$,   0 rad
   $\theta = 33.4°$, 0.582 rad
   $\theta = 53.1°$, 0.927 rad
   $\theta = 90°$,   1.57 rad
2. $\theta = 90°$,   1.570 rad
   $\theta = 72.5°$, 1.266 rad
   $\theta = 25.8°$, 0.451 rad
   $\theta = 0°$,   0 rad
3. $\theta = 45°$,   0.785 rad
   $\theta = 44.7°$, 0.780 rad
   $\theta = 26.6°$, 0.464 rad
   $\theta = 0°$,   0 rad

Exercise 6.13 Trigonometric problems

1.

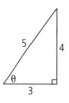

(a) $\sin \theta = \dfrac{4}{5}$, $\cos \theta = \dfrac{3}{5}$, $\tan \theta = \dfrac{4}{3}$ (b) $\theta = 53.1°$

2.

Distance = 2.5 m (2 s.f.)

3.

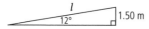

Length = 7.2 m (2 s.f.)

4.

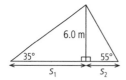

Span = 12.8 m (3 s.f.)

5. $\theta = 33.8°$, $h = 8.22$ m

Exercise 6.14 The parallelogram rule

1. (a) The resultant is 17 N in the original direction.
   (b) The resultant is 7 N in the direction of the 12 N force.
   (c) The resultant is 13 N in a direction at 23° (to nearest °) to the 12 N force.
2. (a) The resultant is 18 N in the original direction.
   (b) The resultant is 3 N in the direction of the 10.5 N force.
   (c) The resultant is 12.9 N in a direction at 36° (nearest °) to the 10.5 N force.
3. The resultant is 2.5 kN in a direction at 37° (nearest °) to the 2.0 kN force.
4. The resultant velocity is 9.4 km/h in a direction at 32° to the 8.0 km/h velocity.
5. (a) The effective velocity is 2.5 km/h in a direction at 53° (nearest °) to the river velocity of 1.5 km/h.
   (b) The time taken is 2.4 minutes.

Exercise 6.15 Resolving vectors

1.

	Horizontal component	Vertical component
(a)	21.7 N	12.5 N
(b)	12.9 N	15.3 N
(c)	11.5 N	8.0 N
(d)	5.0 kN	8.7 kN
(e)	7.1 kN	7.1 kN

2. Component at right angles to the incline = 27.0 N
   Component parallel to the incline = 7.2 N
3. Component at right angles to the incline = 19.6 N
   Component parallel to the incline = 4.2 N
4. Component of velocity in: north direction = 70.7 m/s
   east = 70.7 m/s
   south = 0
   west = 0

5.

	Horizontal component	Vertical component
(a)	173 m/s	100 m/s
(b)	100 m/s	173 m/s
(c)	354 m/s	354 m/s

6. (a) Sum of horizontal components = 7.6 N
   (b) Sum of vertical components = 20.1 N

Exercise 6.16 Phasors

1. (a) V_2 (b) $\pi/2$, 90° (c) 11 V (2 s.f.) (d) 16 V (2 s.f.)
2. Case 1 (a) I leads V by 90° (b) I leads V by $\pi/2$
   Case 2 (a) V leads I by 90° (b) V leads I by $\pi/2$
   Case 3 (a) V leads I by 38° (b) V leads I by 0.66 rad
   Case 4 (a) I leads V by 47° (b) I leads V by 0.82 rad
   Case 5 (a) and (b) I and V are in phase
   Case 6 (a) V_2 leads I by 90°; V_1 is in phase with I (b) V_2 leads I by $\pi/2$
   (c) Case 1 (d) Case 2 (e) Case 5
   (f) $V_{n.m.s.} = 5$ V (g) $V_{peak} = 5\sqrt{2}$ V (h) V leads I by 53° (2 s.f.)

Chapter 7 Answers

Exercise 7.1 Plotting a graph and reading off values

1.

R/Ω	0.5	1.5	2.2	3.2	4.2	5.0	6.0	6.9
V/V	0	2	4	6	8	10	12	14

2. (a) 3.9 m³ (b) 7.1 m³

Exercise 7.2 Plotting points

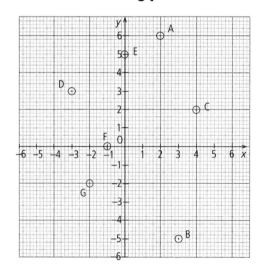

Exercise 7.3 Equations of straight lines

1. L
2. 0
3. K and M
4. M and N
5. 2/3
6. 3
7. $y = 3x$

Exercise 7.4 Drawing straight lines '$y = mx + c$'

1. (a) (i) yes (ii) $y = 6$ (b) (i) $y = -1$ (ii) $y = 5$ (c) (−1, −3)
2. (a) Gradient is 2; y intercept is 3.
   (b)

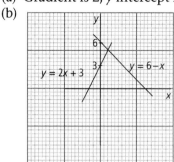

Exercise 7.5 Gradients

1. (a) 5°C per minute. (b) It decreases. (c) The difference between the temperature and that of the surroundings gets less.
2. (a) 0.44 m/s (2 s.f.). (b) Speed. (c) It decreases. (d) The body is decelerating.
3. (a) Rate of rise of temperature. (b) Acceleration. (c) Rate of decay.
   (d) Velocity (speed). (e) Electric current.

Exercise 7.6 Relationships in linear form

1. $v^2 = \dfrac{2hf}{m} - \dfrac{2\phi}{m}$; graph of v^2 against f; gradient $\dfrac{2h}{m}$; intercept $-\dfrac{2\phi}{m}$.

2. $p = \dfrac{k}{V}$; graph of p against $\dfrac{1}{V}$; gradient k; intercept O.

3. $v^2 = 2gh$; graph of v^2 against h; gradient $2g$; intercept O.

4. $R = R_0 + R_0\alpha\theta$; graph of R against θ; gradient $R_0\alpha$; intercept R_0.

5. $\dfrac{x}{v} = av + b$; graph of $\dfrac{x}{v}$ against v; gradient a; intercept b.

Chapter 8 Answers

Exercise 8.1 Graphs of linear relationships

1. (a) Straight line through (0,0).   (b) *Mc*.
2. (a) When $M = 1$ kg, $l = 14$ cm. When $M = 0$, $l = 10$ cm.
   (b) For an accurate graph, l can start at 10 cm, not at zero.
   (c) (i) 0.5 kg   (ii) 12.4 cm   (iii) 3.2 cm
3. (b) Accelerating uniformly.   (c) $8\,\text{m/s}^2$
4. (b) Uniform speed on flat; faster uniform speed downhill.
   (c) 2 m/s   (d) 5 m/s   (e) On the flat.

Exercise 8.2 Conversion graphs

1. (a) 310 g (2 s.f.)   (b) 90 cm³ (2 s.f.)

Exercise 8.3 Miscellaneous experimental graphs

1. (a) $t = \dfrac{1}{10}$ s   (b) $d = 24.3$ mm (3 s.f.)   (c) $v = 243$ mm/s
   (d) $d_B = 69$ mm
   (e)

V/(cm/s)	24.3	15.8	8.2	11.8	20.4	27.4
d_B/cm	6.9	2.8	0.7	1.5	4.6	8.3

   (f) Curve; positive increasing gradient shows that, as speed increases, braking distance goes up more and more.
2. (a) $Q_m = 2.50\,\mu C$ (b) $Q = 1.58\,\mu C$, $\tau = 10\,\mu s$ (2 s.f.)
   (c) $\tau = 10\,\mu s$ (2 s.f.)   (d) $\tau = 10\,\mu s$ (2 s.f.)
3. (b) Uniform acceleration from rest.   (c) 1.8 m/s
   (d) 0.60   (e) $v = 1.8 + 0.60t$.

Exercise 8.4 Gradients of experimental graphs

1. As the temperature rises, the resistance decreases.
   $-1.25\,\Omega$ per °C
2. (a) 13.5 hours   (b) 11 g/h
3. (b) -0.75 A/s
   (c) The current is decreasing at a rate of 0.75 A per second instantaneously after 0.1 seconds from the start.

4. (b) V increases as I increases. Ohm's law applies at constant temperature. Here, the resistance rises as current increases because of heating.
   (c) 800 V/A (2 s.f.)

Exercise 8.5 Graphs on the same axes

1. (a) Acceleration when slope is at angle θ_1.
   (b) Gradient increases because acceleration is greater.
   (c) 10m/s^2 (d) 90°

Exercise 8.6 Working with a number of variables

1. (a) ρ and a.   (b) Straight line through (0,0), positive gradient.
   (c) ρ/a (d) $\rho = \dfrac{\pi d^2}{4} \times$ gradient.

2. (a) l, b and d.
   (b) y against w, straight line through (0,0), gradient giving $\dfrac{kl^3}{bd^3}$.
   (c) Vary l with w constant; plot y against l^3; straight line through (0,0), gradient giving $\dfrac{kw}{bd^3}$.

Exercise 8.7 Reducing to linear form

1. (a) $\dfrac{x_s}{v} = av + b$
   (b) Graph of $\dfrac{x_s}{v}$ against v; $a = 0.018\,\dfrac{\text{yd}}{(\text{m.p.h.})^2}$ (2 s.f.); $b = 0.32\,\dfrac{\text{yd}}{\text{m.p.h.}}$ (2 s.f.)
   (c) $a = 0.082$ kg/N (2 s.f.); $b = 0.65$ s (2 s.f.)
   (d) b to reaction time; a to braking force.
   (e) 110 yd at 70 m.p.h.
   (f) Stopping distance is influenced by both reaction time and braking force but is much greater at high speeds because of the av^2 term.

Exercise 8.8 Area under a graph

1. (a) 0.50 m/s^2 (2 s.f.)   (b) 0.25 m/s^2 (2 s.f.)
   (c) 30 m/s   (d) 9 minutes 10 seconds   (e) 13.8 km
2. (a) 0.95 J   (b) Less   (c) The rubber gets hot.

Exercise 8.9 Applications of area under a graph

1. 2.5 µC (2 s.f.)
2. (b) 0.4 mA   (c) 1 µC   (d) 0.4 mA

Answers to self-assessment test

The correct responses to the multiple-choice self-assessment test are:

1. c	11. d	21. d	31. b	41. a
2. c	12. a	22. a	32. c	42. b
3. a	13. d	23. b	33. b	43. c
4. b	14. a	24. d	34. c	44. c
5. c	15. b	25. c	35. b	45. d
6. c	16. c	26. b	36. c	46. c
7. d	17. a	27. c	37. a	47. a
8. a	18. a	28. d	38. c	48. c
9. b	19. b	29. c	39. c	49. b
10. d	20. c	30. b	40. b	50. b

The SI system of units

In 1960 a new system of units was introduced that rapidly gained recognition world-wide. This system is known as the SI system, which is an abbreviation of the International System of Units (Système international d'unités), and makes use of seven base units. These are listed below, each with its unit and unit symbol.

SI base units and symbols

Quantity	Name of unit	Symbol
Length	metre	m
Mass	kilogram	kg
Time	second	s
Electric current	ampère	A
Thermodynamic temperature	kelvin	K
Luminous intensity	candela	cd
Amount of substance	mole	mol

Engineering science often involves quantities that are very much larger or smaller than a given base unit. Prefixes giving powers of ten which are a multiple of three are often used. The standard prefixes most commonly used are listed below.

SI unit prefixes

Multiple or submultiple	Prefix	Symbol
$1\,000\,000\,000\,000 = 10^{12}$	tera	T
$1\,000\,000\,000 = 10^{9}$	giga	G
$1\,000\,000 = 10^{6}$	mega	M
$1000 = 10^{3}$	kilo	k
$100 = 10^{2}$	hecto	h
$10 = 10^{1}$	deca	da
1 unit		
$0.1 = 10^{-1}$	deci	d
$0.01 = 10^{-2}$	centi	c
$0.001 = 10^{-3}$	milli	m
$0.000\,001 = 10^{-6}$	micro	μ
$0.000\,000\,001 = 10^{-9}$	nano	n
$0.000\,000\,000\,001 = 10^{-12}$	pico	p

From the base units, the units of all other quantities are derived. The table gives the base and derived units for the quantities you meet in this book, together with the correct abbreviations for the units and the symbols for quantities you will find in the text.

Quantities, symbols, units and abbreviations

Quantity	Usual symbol	Unit	Abbreviated unit
Length	l	metre	m
Area	A	metre^2	m^2
Volume	V	metre^3	m^3
Mass	m	kilogram	kg
Time	t	second	s
Density	ρ	kilogram/metre^3	kg m^{-3} or kg/m^3
Speed	v	metre/second	m s^{-1} or m/s
Acceleration	a	metre/second^2	m s^{-2} or m/s^2
Force	F	newton	N
Work	–	joule	J
Energy	–	joule	J
Power	–	watt	W
Weight	W	newton	N
Pressure	p	pascal	Pa
Moment	–	newton metre	N m
Torque	Γ	newton metre	N m
Temperature	T	kelvin	K
Frequency	f	hertz	Hz
Electric current	I	ampère	A
Voltage	V	volt	V
Resistance	R	ohm	Ω
Charge	Q	coulomb	C

NOTE: There are no symbols consistently used for the quantities work, energy, power and moment of a force.

In engineering practice many other units are used but conversion from one system to another is beyond the scope of this book. The unit of temperature 'degree Celsius' is however in common use. Its relationship with the kelvin is

$$T/K = \theta/°C + 273$$

where T = kelvin temperature
θ = temperature in degrees Celsius

Index